Ergonomics - New Insights

Edited by Orhan Korhan

Published in London, United Kingdom

Ergonomics - New Insights
http://dx.doi.org/10.5772/intechopen.102238
Edited by Orhan Korhan

Contributors
Patience Erick, Tshephang Tumoyagae, Tiny Masupe, Daniel Oluwafemi Odebiyi, Udoka Arinze Chris Okafor, Özlem Kaya, Heather Carnahan, Emily Walsh, Orhan Korhan, Muhamed Fallaha, Zeki Murat Çınar, Qasim Zeeshan

First published in London, United Kingdom, 2023 by IntechOpen
IntechOpen is the global imprint of INTECHOPEN LIMITED, registered in England and Wales, registration number: 11086078, 5 Princes Gate Court, London, SW7 2QJ, United Kingdom

British Library Cataloguing-in-Publication Data
A catalogue record for this book is available from the British Library

Additional hard and PDF copies can be obtained from orders@intechopen.com

Ergonomics - New Insights
Edited by Orhan Korhan
p. cm.
Print ISBN 978-1-80356-470-8
Online ISBN 978-1-80356-471-5
eBook (PDF) ISBN 978-1-80356-472-2

Meet the editor

Orhan Korhan graduated with a BS from Eastern Mediterranean University (EMU), Northern Cyprus, in 2000, an MS from the University of Louisville, USA, in 2002, and a Ph.D. in Industrial Engineering from EMU in 2010. He has been working at EMU since 2009. He became an assistant professor in 2010, an associate professor in 2014, and a full professor in 2020. He is assigned to scientific committees at several international conferences and has published several books, book chapters, and papers. His current research interests include work-related musculoskeletal disorders, cognitive ergonomics, Industry 4.0, and facilities planning and design.

Contents

Preface

Humans are the most crucial elements of any production system, regardless of any relevant technological advancements. As such, the main goal of any system is to design the workspace in a way that maximizes the productivity of the human operator. When attempting to boost output, effectiveness, and quality within the production system, ergonomics is the most important science to consider.

Ergonomics has three pillars: physical, organizational, and cognitive. This book presents recent advances in each of these three pillars to guide researchers in future studies in ergonomics. It is organized into four sections.

The first section introduces the topic. The second section presents current studies on physical ergonomics, including studies on musculoskeletal disorders, injuries, and prevention strategies. The third section focuses on organizational ergonomics and examines work design, working conditions, and workplace design. The fourth and final section examines cognitive ergonomics and the role of the human operator in the era of Industry 4.0.

I would like to thank Ms. Sara Tikel who helped me with a positive attitude at every single step of the publication process. Moreover, I would like to thank to my son, Kaya, for the love and inspiration he brought to our lives.

Orhan Korhan
Department of Industrial Engineering,
Eastern Mediterranean University,
Famagusta, North Cyprus

Section 1

Introduction

Chapter 1

Introductory Chapter: Contemporary Topics in Ergonomics

Orhan Korhan

1. Introduction

The human body is physically and physiologically limited. Also, human brain has boundaries. Therefore, the human is capable of implementing certain abilities, both physically and intellectually. In order to minimize the limitations of the human and maximize its efficiency in the system it exists, its interactions with the elements of the system should be investigated and understood carefully.

Ergonomics is a multidisciplinary science that optimizes human well-being and overall system performance by using theory, principles, data, and methods.

The term ergonomics refers to the science of work and is derived from the Greek ergon (work) and nomos (laws). Ergonomics is also known as "human factors," and both terms can be used together or interchangeably [1].

Ergonomics is therefore regarded as a science and a profession at the same time. It is the study of how humans interact with other system components. It also seeks to maximize human well-being and system performance through the application of facts, principles, theories, and methods to design concepts.

The study of "physical, cognitive, sociotechnical, organizational, environmental, and other important elements, as well as the complex interactions between humans and other humans, the environment, tools, goods, equipment, and technology" is known as ergonomics [1]. Thus, it can be subdivided further into physical, cognitive, and organizational ergonomics to assess their individual and combined effects on people working within a given system.

In summary, physical ergonomics deals with the human physical body, cognitive ergonomics with the human brain, and organizational ergonomics with systems and the cultures that exist within them.

This chapter is designed to discuss each type of ergonomics in detail, indicate the recent advancements, and propose future studies that can be conducted in these fields in the last section.

2. Physical ergonomics

The study of the human body's responses to physical and physiological job demands is known as physical ergonomics. It studies how anatomy, anthropometry, biomechanics, physiology, and the physical environment influence physical activity [2].

IntechOpen

Physical ergonomics addresses the effects of repetitive motion, materials handling, workplace safety, and comfort in the use of portable devices, keyboard design, working postures, and the work environment. The most common types of difficulties are repetitive strain injuries caused by repetition, vibration, force, and posture and thus have design implications [3]. Thus, the main research areas of physical ergonomics include repetitive movement, work-related musculoskeletal disorders, health, working postures, workplace layout, safety, equipment design, and material handling.

Studies have indicated that when workers have less physical strain and difficulties completing their jobs, they are more productive. Moreover, quality and profitability can be directly harmed by poor ergonomics. Errors and decreased work quality can result from physically and mentally exhausted users of ergonomically flawed designs [4, 5].

3. Organizational ergonomics

Organizational ergonomics, which is also known as macroergonomics, examines the interaction between systems and organizations interact, as well as designs of the systems. It entails knowing how to improve work systems in order to improve overall performance and effectiveness of an organization. Thus, it entails optimizing an organization's sociotechnical system's policies, processes, and structures.

Within an organization, organizational ergonomics can be approached in a variety of ways, including top-down, bottom-up, and middle-out. Those in leadership or management positions may recommend the work flow, structure, and resources available to perform work in order to improve organizational ergonomics using a top-down approach. A middle-out strategy entails investigating an organization's internal workings to ascertain the efficiency of its work systems and procedures at all levels of the "organizational hierarchy." When using a bottom-up strategy, employees must actively participate and contribute in order to identify issues and potential fixes [6].

To enhance organizational ergonomics in a business, it may be necessary to assess certain workplace factors, including; work design, virtual organizations, teamwork, quality management, telework or remote work, participatory design, new work paradigms, design of work times, crew resource management, cooperative work, community ergonomics, and communication [7, 8].

4. Cognitive ergonomics

The field of cognitive ergonomics examines how the human brain interacts with and processes information, and how well a person performs within a particular system as a result.

This branch of cognitive ergonomics studies "mental workload," which includes training and decision-making. Mental functions like perception, attention, memory, reasoning, making decisions, learning, and motor response are taken into account because they have an impact on how people interact with other mechanical components of a system. The ways in which a person interacts with and completes their work are influenced by perception, memory, reasoning, and motor response. The worker feels more stress as a result of a heavier cognitive workload [9].

Ergonomists in this field evaluate and offer recommendations on education, design, usability, skill training, physical training, human-technology interaction, work stress, decision-making processes, social stress, and fatigue in addition to mental workload.

5. Discussion and conclusion

The human remains the most important component of any production system despite of the technological advances. Gaining knowledge of the three main sorts can help us better understand the complicated workplace found on today's sophisticated manufacturing floors.

In today's technologically competitive production environment, physical ergonomics is evolving into automation of the repetitive manual tasks, and manual handling involving logistics and transportation are being improved by new digital technologies such as autonomous robots. Even though ergonomic response is getting better with wearable and handheld devices, there are still risks arise from close human-machine collaboration.

Organizational ergonomics, on the other hand, is moving toward studying the requirements of hybrid production systems where humans and machines are becoming more integrated. Thus, the design and organization of work will be impacted by this new human-machine interaction. It is expected that employers will gain from human-centered design. So, the interaction of the new technologies and work organization will determine how future skills of the human operators to be developed.

The most important aspect of ergonomics will be the cognitive ergonomics in this new technology dominated era. Virtual models facilitate timely interactions and enhance perception, cyber physical systems are developing new ways for people to interact with machines, the use of augmented reality tools will lessen mental stress, and data exchange between departments is expected to enhance cognitive ergonomics. Thus, IT and problem-solving abilities will be unavoidably required.

Therefore, each element of organizational, cognitive, and physical ergonomics can be used separately or more effectively in combination with one another. Even though each of these lists might seem overwhelming in its entirety, rest assured that focusing on even one will help increasing the efficiency.

Author details

Orhan Korhan
Eastern Mediterranean University, Famagusta, North Cyprus

*Address all correspondence to: orhan.korhan@emu.edu.tr

References

[1] IEA - International Ergonomics Association. What is Ergonomics? Available at: https://iea.cc/what-is-ergonomics/. [Accessed: September 6, 2022]

[2] Helander M. A Guide to Human Factors and Ergonomics. Second ed. Boca Raton, FL, USA: CRC Press; 2005

[3] Karwowski W, editor. Handbook of Standards and Guidelines in Ergonomics and Human Factors. Lawrence Erlbaum Associates. Boca Raton, FL, USA: CRC Press; 2005

[4] Salvendy G. Handbook of Human Factors and Ergonomics. Hoboken, New Jersey, USA: John Wiley & Sons, Inc; 2012

[5] Woodson WE. Human Factors Design Handbook: Information and Guidelines for the Design of Systems, Facilities, Equipment, and Products for Human Use. New York, USA: McGraw-Hill; 1992

[6] Becker T, Stern H. Future trends in human work area design for cyber-physical production systems. In: Procedia CIRP – 49th CIRP Conference on Manufacturing Systems (CIRP-CMS 2016). Vol. 57. 2016. pp. 404-409. DOI: 10.1016/j.procir.2016.11.070

[7] Dworschak B, Zaiser H. Competences for cyber-physical systems in manufacturing – first findings and scenarios. In: Procedia CIRP - 8th International Conference on Digital Enterprise Technology – DET 2014 – "Disruptive Innovation in Manufacturing Engineering Towards the 4th Industrial Revolution Competences. Vol. 25(C). 2014. pp. 345-350. DOI: 10.1016/j.procir.2014.10.048

[8] Müller U, Gust P, Feller N, Schiffmann M. WorkDesigner: Consulting application software for the strain-based staffing and design of work processes. In: Procedia Manufacturing – 6th International Conference on Applied Human Factors and Ergonomics (AHFE 2015) and the Affiliated Conferences, AHFE 2015. Vol. 3. 2015. pp. 379-386. DOI: 10.1016/j.promfg.2015.07.179

[9] Vernim S, Walzel H, Knoll A, Reinhart G. Towards capability-based worker modelling in a smart factory. In: 2017 IEEE International Conference on Industrial Engineering and Engineering Management (IEEM). Vol. December. Singapore: IEEE; 2017. pp. 1576-1580. DOI: 10.1109/IEEM.2017.8290158

Section 2

Physical Ergonomics

Chapter 2

Musculoskeletal Disorders, Workplace Ergonomics and Injury Prevention

Daniel O. Odebiyi and Udoka Arinze Chris Okafor

Abstract

Musculoskeletal Disorders (MSDs) affect body parts, with severity ranging from mild to intense. When MSDs develop in occupational settings, sequel to the physical tasks involved in the performance of work and the condition of the work-environment, they are referred to as work-related musculoskeletal disorders (WMSDs). The development and prognosis of any particular MSDs are modified by multiple risk factors, which are physical, individual, and psychosocial, in nature. None of these factors act separately to cause WMSDs, rather, they interact. The goal of ergonomics is to create an ergonomically sound work-environment, with the view to reducing the occurrence of WMSDs. This is premised on adherence to effective workplace ergonomic principles (WEP). By and large, WEP is more effective when done both at the workplace and during the performance of leisure time activities. Often, WEP involves designing the workplace, with consideration for the capabilities and limitations of the workers, thus promoting good musculoskeletal health, and improving performance and productivity. For favorable outcomes, a three-tier hierarchy of controls (Engineering, Administrative, and use of Personal Protective Equipment) is widely accepted as a standard intervention strategy for reducing, eliminating, or controlling workplace hazards. Failure of this strategy will expose workers to WMSDs.

Keywords: musculoskeletal disorders, workplace, ergonomics principles, injury prevention

1. Introduction

Musculoskeletal disorders (MSDs) is used to described injuries or disorders of the musculoskeletal system, like muscles, nerves, tendons, ligaments, joints, and cartilage; including the supporting structures of neck and back, and can affect all parts of the body (**Table 1**). Musculoskeletal disorders are described as Work-related, i.e. Work-related Musculoskeletal Disorders (WMSDs) when they are caused, and/or made worse or persists longer than expected, by the performance of work/task, vis-a-vis., work-environment and work-conditions [1]. According to the Bureau of Labor Statistics (BLS), MSDs represent one of the largest work-related problems in the United States; with the incidence rate higher among male full-time workers compared with females [2]. And according to the 2020/21 Labour Force Survey (LFS) of the United Kingdom,

Serial No.	Structures of the Musculoskeletal system	Examples
1	Muscle/Tendon (Including inflammation of the tendons and/or their synovial sheaths)	Muscle Sprain/Strain Muscles fatigue eg Tension Neck Syndrome Rotator Cuff Tendonitis Epicondylitis, Bursitis Tendon strain
2	Nerve (usually involve the compression of nerve)	Numbness/Tingling Digital Neuritis Radial Tunnel Syndrome Trigger Finger Carpal Tunnel Syndrome
3	Joint	Joint Pain Proprioception Tear Temporomandibular joint (TMJ) Pain
4	Ligament	Ligament Sprain
5	Cartilage	KOA
6	Vascular (Affectations of the blood vessels)	vibration syndrome
7	Supporting structures of neck and back	Mechanical Back Pain Stress Anxiety

Table 1.
Common musculoskeletal disorders.

470,000 workers are suffering from WMSDs - new or long-standing [3]; the occurrence and pattern of WMSDs in the United Kingdom are as shown in **Figure 1**. In Nigeria, WMSDs are especially prevalent in certain occupational sectors and industries such as transportation, warehousing, manufacturing/petroleum industry, health care, Communication services, Butchers, agriculture, and construction services [4–9].

Work-related Musculoskeletal Disorders are classified according to the affected musculoskeletal/anatomical structure (**Table 1**). According to the Bureau of Labor Statistics (BLS), MSDs accounted for 32% of all injury and illness cases in 2014 among full-time workers [2]. Work-related musculoskeletal disorders usually develop over time, in form of cumulative micro-traumas, sustained while working; it development can also be episodic. Additionally, the severity can progress from mild (i.e., Occasional) to severe/intense (i.e., chronic). These disorders are seldom life-threatening but they impair the quality of life of a large proportion of the adult working population.

The National Institute for Occupational Safety and Health (NIOSH) defined WMSDs as those diseases and injuries affecting the musculoskeletal, peripheral nervous, and neurovascular systems and are caused or aggravated by occupational exposure to ergonomic hazards [10]. Ergonomic hazards refer to physical stressors and workplace conditions that pose a risk of injury or illness to workers' musculoskeletal system. Ergonomic injury risks include repetitiveness and pace of work (i.e., repetitive motions), forceful motions, vibration, extreme temperatures (especially cold conditions), awkward work-posture and movements, caused by the inadequate design of work-stations, tools or other work equipment, and by improper work methods [11]. Other risk factors include, lack of influence or control over one's job, increase pressure (e.g., to produce more), lack of or poor communication, monotonous tasks,

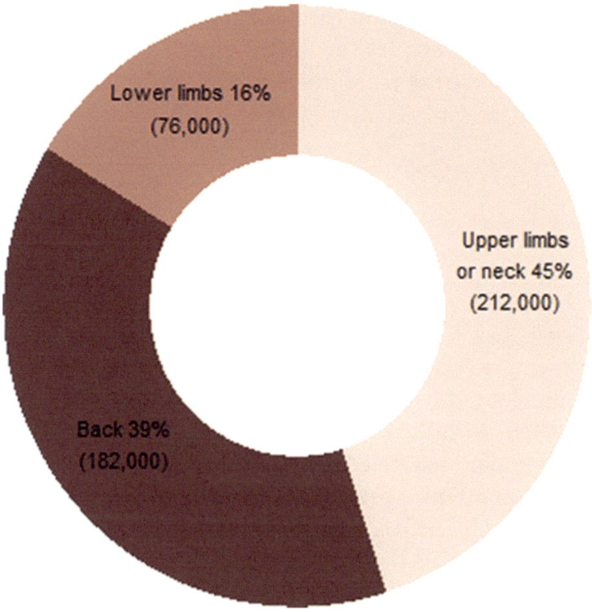

Figure 1.
Occurrence and pattern of work-related musculoskeletal disorders [3].

and perception of low support (e.g., management or co-workers). Furthermore, MSDs have been reported to be associated with reduced work ability, and decreased productivity among workers, across working populations [4, 7–9, 12–14]. According to the Bureau of Labour Statistics of the Department of Labour, MSDs is the diseases and/or disorders of the musculoskeletal system, and connective tissue, when the event or exposure leading to the case is bodily reaction (e.g., bending, climbing, crawling, reaching, twisting), overexertion, or repetitive motion [15]. They are not the result of any instantaneous non-exertion event like slips, trips, falls or similar incidents.

The occurrence of WMSDs has been attributed to the exposure of workers (employees) to physical factors at work, as a result of poor and/or none adherence to Work-place Ergonomic Principles (WEP). The occurrence of WMSDs are basically attributed to the performance of work, and work-environment, furthermore, MSDs are usually made worse and/or longer lasting by work conditions that preclude good ergonomic principles during execution. Therefore, adherence to WEP is essential in preventing the occurrence of WMSDs. Work-place ergonomic principles involves identifying, analyzing, and controlling work-place risk factors, for the purpose of preventing and/or reducing the occurrence of MSDs (i.e. soft tissue injuries), caused by performance of work, vis-a-vis, exposure to sudden or sustained force, vibration, repetitive motion, and awkward posture etc. This is achieved by creating an ergonomically sound work-environment. By and large, adherence to effective WEP helps to create a workplace condition and job demands, that is, at the capacity of the workers (working population), and thus can be very helpful in preventing/reducing WMSDs. Work-place ergonomic principles is particularly recommended in the conduct of all type job descriptions; including materials manual handling, office-work, and patients management, and rehabilitation.

This book chapter is based on detailed literature of the causes and prevention of WMSDs. It described the inherent workplace hazards, workers are exposed to, at the different work-environments; and the benefits of effective preventive strategies, using standard ergonomic principles (i.e., WEP), applicable for materials manual handling work-place, and during patients management/rehabilitation.

2. Musculoskeletal disorders – causes and prevention

2.1 Background

The disorders/injuries of the soft tissues of the musculoskeletal system - muscles, nerves, tendons, ligaments, joints, and cartilage, are commonly referred to by many names, including musculoskeletal disorders (MSDs), repetitive strain injuries (RSI), repetitive motion injuries (RMI), cumulative trauma disorders (CTDs) and overuse injury [16]. The problem with using other terminology other than MSDs, is that they appear to suggest a singular causative factor (e.g. repetition or stress) as the cause of the soft tissue disorders. This is restrictive, because the literature points to multiple causative risk factors for MSDs. The World Health Organization (WHO), has reported that WMSDs has multi-factorial aetiology; indicating that a number of risk factors contribute to causing these disorders [17, 18]. These factors are physical, work organizational, work-environment, work-conditions (i.e. repetition, sudden/forceful exertions - like lifting a heavy object, and repetitive/sustained awkward postures), psycho-social, individual, and sociocultural, in nature [17]. This multi-factorial aetiology is the major reason for the controversy surrounding WMSDs – as both multiple and individual factors have been identified in the development of WMSDs [17, 18]. The development of MSDs has been recognized as having occupational aetiology factors as early as the beginning of the 18th century [1]. However, it was not until the 1970s that occupational factors were examined using epidemiologic methods, and the work-relatedness of these conditions began appearing regularly in the international scientific literature [1].

2.2 Causes of musculoskeletal disorders

The musculoskeletal system (i.e., muscles, nerves, tendons, ligaments, joints, and cartilage) are most effectively utilized, when they are exposed to little or no work-place risk; i.e. one that is within the worker's capability. The level of risk depends on the intensity, frequency, and duration of the exposure to these work-place risks/hazards. Furthermore, the effects of work-place risks may be amplified by organizational factors such as shift work, work pace, imbalanced work-rest ratios, demanding work standards, lack of task variety etc. Subjecting a worker to work (carry out a task), in an ineffective WEP, is making the worker to work outside his/her body's capabilities and/or limitations. Simply put, the worker is being asked to put his/her musculoskeletal system at risk. This may lead to body fatigue in the workers, beyond their ability to recover, and which may result in musculoskeletal imbalance, and may eventually, lead to the development of MSDs . Thus, exposure to work-place risk factors, as a result of ineffective WEP, puts workers at risk of developing MSDs (**Figure 2**). According to the Bureau of Labour Statistics of the Department of Labor, the diseases and/or disorders of the musculoskeletal system (and connective tissue) is described, when the event or exposure leading to the case is bodily reaction (e.g., bending, climbing, crawling, reaching, twisting), overexertion, or repetitive motion [15]. As a matter of facts, "... there is an international near-consensus that MSDs are causally related to occupational

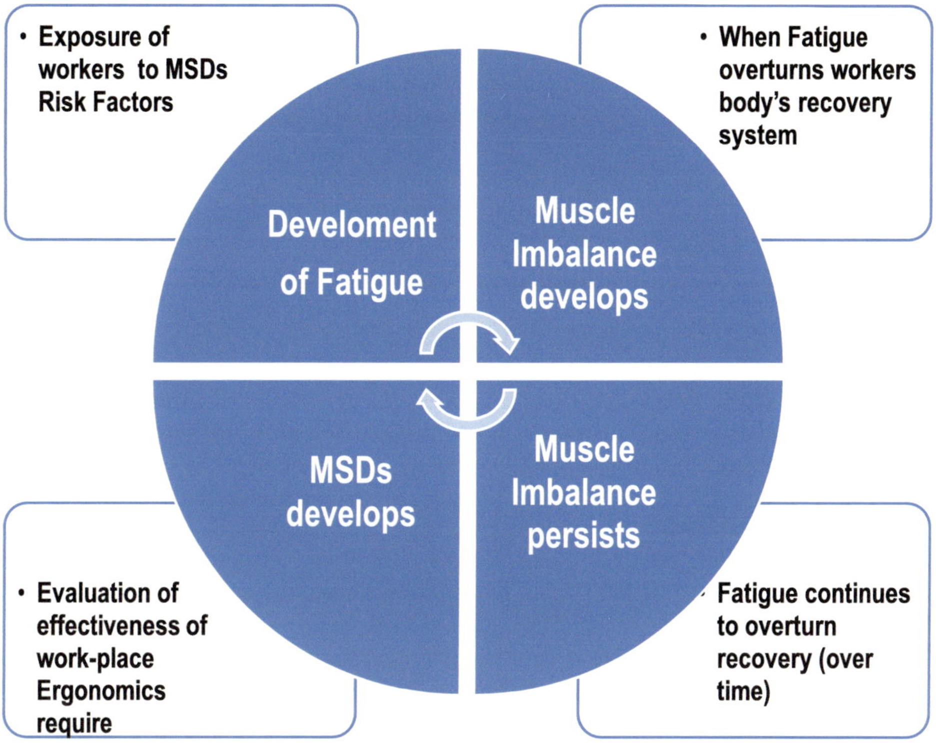

Figure 2.
Mechanism development of MSDs.

ergonomic stressors, such as repetitive and stereotyped motions, forceful exertions, non-neutral postures, vibration, and combinations of these exposures" [19].

Work-related Musculoskeletal disorders do not result from instantaneous non-exertion events like slips, trips, falls etc. The main cause of WMSDs is exposure to (ergonomic) "Risk Factors" at the work-place. Thus, the disposition for developing WMSDs is related more to the difference between the demands of work and the worker's physical work capacity, which decreases with age [20]. Therefore, adherence to an effective Work-place ergonomic principle/design is essential in the prevention of the development of MSDs. Epidemiological studies [1, 21] have categorized work-place ergonomic risk factors into three, namely: (a) Physical factors (like sustained or awkward postures, repetition of the same movements, forceful exertions, hand-arm vibration, all-body vibration, mechanical compression, and cold) (b). Individual factors (like age, gender, professional activities, sport activities, domestic activities, recreational activities, alcohol/tobacco consumption and, previous WMSDs), (c). Psychosocial factors (like work-pace, autonomy, monotony, work/rest cycle, task demands, social support from colleagues and management and job uncertainty). In this book chapter, ergonomic risk factors are broadly divided into two categories: work-related (ergonomic) risk factors and individual-related risk factors.

2.2.1 Work-related risk factors

This is further divided into two - Primary and secondary factors. There are three primary work-related (ergonomic) risk factors, which are basically physical in nature:

i. **Sudden/Forceful Exertions:** These are work-tasks and cycles, that require high force loads on the human body, like heavy lifting, pulling, pushing a heavy objects, or excessively squeezing a hand tool such as a hammer. During execution by the workers, muscle effort increases, in response to the high force requirements, thus, leading to increased associated fatigue, and subsequent musculoskeletal imbalance, particularly when fatigue overturns workers body's recovery system. And over time, this will eventually leads to the development of MSDs, as fatigue continues to overturn recovery. A study by a group of Swedish researchers demonstrated, using Doppler ultrasound scans, that chronic (prolonged) contraction of a muscle can cause a narrowing of the blood supply to the muscle due to compressive effect on the muscle [22]; thus reducing circulation to the muscle fibers and increasing the time required for recovery. This can be precipitated by both static and dynamic muscular contraction (loading) and duration, although static loading is a greater risk factor than dynamic loading, since static loading results in increased muscle fiber recruitment and fatigue and decreased blood perfusion. Forceful exertions produce increased muscle effort in response to high task load, thus, leading to more rapid muscle fatigue and overuse which can lead to upper extremity injuries.

ii. **High (Task) Repetitions:** These are work-tasks and cycles that are repetitive in nature, that is, require making the same motions repeatedly. They are frequently controlled by hourly or daily production targets and work processes. A task is considered to be highly repetitive if the cycle time is 30 seconds, or less, or if a task or motion is performed more than 50% of the time it takes to complete the work cycle [23]. Work and rest cycles are the intervals of time measured during one complete task revolution or cycle. The more repetitive the task or cycle is, the less recovery time there is for the muscles and tendons. Inappropriate rest/work cycles are work cycles that do not allow time for sufficient recovery; this may lead to the accumulation of micro trauma, sequel to exposure to ergonomic hazard. Thus, leading to CTDs. When combined with other risks factors (e.g. sudden/forceful exertion and/or awkward postures), high repetition tasks can lead to increased fatigue, and subsequent musculoskeletal imbalance, particularly when the fatigue overturns workers body's recovery system. And will eventually leads to the development of MSDs, as fatigue continues to overturn recovery, over time.

iii. **Awkward postures (Repetitive or Sustained):** The positions of the wrist and arm are often considered during awkward postures while executing out the tasks, which may be repetitive or sustained. Awkward postures are those in which joints are held or moved away from the body's natural position; like prolonged standing/sitting, significant sideways twisting, reaching above shoulder height, one handed lifting/carrying, kneeling and squatting (**Figure 3**). The closer the joint is to its end of range of motion, the greater the stress that is placed on the soft tissues of that joint (muscles, nerves, and tendons). The joints of the body are most efficient when they operate closest to the mid-range motion of the joint. Risk of MSDs is increased when joints are worked outside of this mid-range repetitively or for sustained periods of time without allowing for adequate recovery time. Assumption of awkward postures creates an ergonomic hazard, as it place excessive stress/force (overload) on the musculoskeletal structures (i.e. muscle, joints, cartilage, and tendons) in an asymmetrical manner, thereby imposing a static load, and thus reduces nerve and muscle blood flow on these structures. An example is any activity that uses repetitive finger

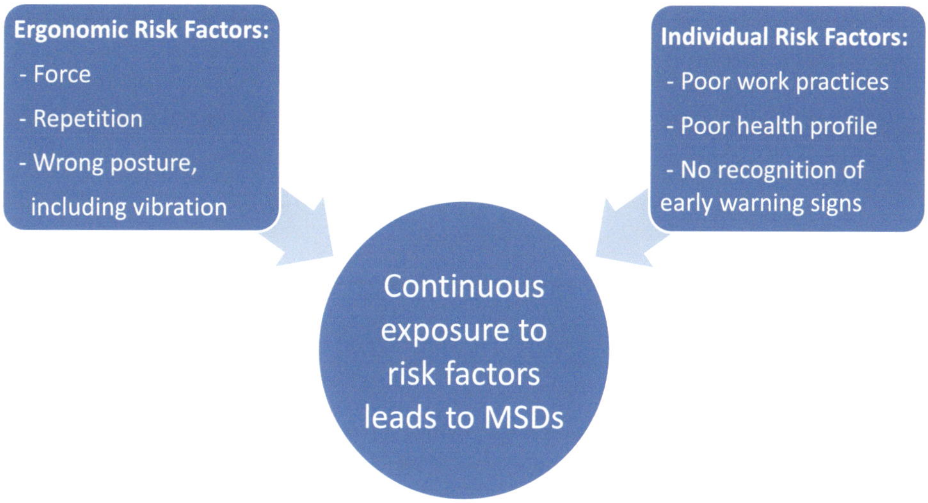

Figure 3.
Work-related musculoskeletal disorders risk factors.

motions with the wrist in an extended position, and in a constrained postures, such as playing a musical instrument or typing. Exposure to this risk factor, in combination with other risks factors like repeated exposure to force, vibration, awkward posture or repetitive lifting of heavy objects in extreme or awkward postures, can lead to increased fatigue, and subsequent musculoskeletal imbalance. This will eventually leads to the development of MSDs, as fatigue continues to overturn recovery, over time. For instance, combined exposure to prolonged sitting in awkward postures may increase the risk. Exposure to these workplace risk factors puts workers at a higher level of MSD risk. That is, high task repetition, forceful exertions and repetitive/sustained awkward postures, fatigue the worker's body beyond their ability to recover, leading to a musculoskeletal imbalance and eventually an MSD.

NB: The risk of developing MSDs increases with increasing number of (ergonomic) risk factors involved/present in the execution of the job-task. Jobs that combine high force and high repetition, in awkward postures, pose the greatest risk.

Secondary Risk Factors: These include psychosocial factors [like work-pace, autonomy, monotony, work/rest cycle, task demands, social support from colleagues and management and job uncertainty]. Other secondary risk factors include: Static Posture, Contact Stress, extreme temperature [Cold/heat], Vibration, Noise, Physical Stress, Emotional Stress.

2.2.2 Individual-related risk factors

When all work-related (ergonomic) risk factors are addressed, it is imperative that consideration be giving to individual risk factors, in addition to work-related (ergonomic) risk factors; more so because human beings, are multi-dimensional in nature. This is because limitation to a singular cause of MSDs will limit the ability to create a prevention strategy that addresses the multi-dimensional nature of the workers (and their work-environments). Individual risk factors include:

i. *Poor Work Practices:* Workers should be familiar with the appropriate work practice, using appropriate training strategies, most especially at the entry point. For instance, proper work - practices, body mechanics and lifting techniques, will help avoid unnecessary risk factors capable of leading to the development of MSDs.

ii. *Poor Overall Health Habits:* Workers who exhibit certain poor health habits are at risk of development of MSDs. Poor health status has been reported to increase the effect of exposure to ergonomic hazard, on the musculoskeletal system.

iii. *Poor Rest and Recovery*: Musculoskeletal disorders develop when fatigue outruns the workers recovery system, causing a musculoskeletal imbalance. Workers who do not get adequate rest and recovery put themselves at higher risk.

iv. *Poor Nutrition, Fitness and Hydration*: Workers who do not take care of their bodies are putting themselves at a higher risk of developing MSDs. Also, selection of workers to task station should be based on the capacity of the workers - such that there is no mismatch between the physical fitness level of the workers and the assigned task.

NB: It is imperative to note that, in other to ensure proper balance of work-practice, and subsequently reduce the occurrence of MSDs, both work-related (ergonomic) risk factors, and individual-related risk factors should be adequately evaluated and controlled. And in addition to the adequate control of work-/individual-related risk factors, there is also need for the workers to exhibit: proper work-practice, good health habits, adequate rest (that allows adequate recovery), and a good nutrition, and fitness regimen, otherwise, they will be at greater risk for fatigue, which may outrun their recovery system. Also, having a poor overall health profile, may put workers at greater risk of developing musculoskeletal imbalance and eventually MSDs.

2.3 Prevention of musculoskeletal disorders

Truly, work-place injuries are not inevitable. Therefore, a work-place design plays a crucial role in reducing the development of MSDs in a work-place; and this can be achieved through the application of an effective work-place ergonomic principles (WEP). The main goal of an effective workplace ergonomic principles WEP is to develop or modify work-environment to meet workers' needs. The design of WEP is directed towards improving ergonomic risk factors in the work-place, following a proper ergonomic evaluation of the workstation design, worker's capabilities, workers' physical attributes and habits. The development and implementation of work-place ergonomic controls is based on the correct assessment of the ergonomic risks inhering in the execution of task, with the designs directed to reducing these risk factors.

2.4 Developing and implementing workplace controls

An adequate "prevention strategy", with better outcomes is usually conducted following a holistic evaluation of the work-environment, work-task, and the worker. The evaluation of any work-task (Job), vis-a-vis., workplace (including Tools), Tasks and workers, should necessarily involved identifying Ergonomic Stressors, like the:

i. *Force required to complete the task* - whether the completion/execution of the task involves assumption of static-working or awkward postures.

ii. *Repetitiveness of a task* - whether the work-task is repetitive in nature, that is, require making the same motions repeatedly, for at least 30 seconds, or when a task or motion is performed more than 50% of the time it takes to complete the work cycle;

iii. *Quality of the Task-posture* - whether the Task-posture is prolonged, i.e., more than one hour, or whether the posture is considered awkward or not. Awkward postures place excessive force on joints and overload the muscles and tendons around the effected joint.

iv. *Quality of the worker's rest period* - whether the pace of work (task) allows sufficient recovery between task-movements. Workers need an acclimatization period to become accustomed to their work demands and to be considered work-hardened or task-fit, this is particularly needed if the workload or working conditions are changed, and/or following an extended absence from work.

v. *Overall health of the workers* - whether the workers exhibit adequate health, including: proper work-practice, good health habits, adequate rest (that allows adequate recovery), and a good nutrition and fitness regimen. It is important to note that, combination of postures, forces and frequencies, increase the chance of developing an MSD.

For adequate outcomes, a three-tier hierarchy of controls (of Engineering, Administrative and use of Personal Protective Equipment), is widely accepted as an intervention strategy for reducing, eliminating, or controlling workplace hazards, including ergonomic hazards. These are:

2.4.1 Engineering controls

This entails, designing the job-task, to take account of the capabilities and limitations of the workers using engineering controls. Engineering improvements/controls include:

i. Rearranging, modifying, redesigning, or replacing the work-station/task-process.

ii. Changing the task-process, vis-a-vis., using handles or slotted hand holes in packages requiring manual handling, or changing the way materials (parts/products) are transported - using mechanical assistive devices to relieve heavy load lifting and carrying tasks.

iii. Changing workstation layout, including using height-adjustable workbenches or locating tools and materials within short reaching distances - For example, a job that requires sitting for long periods of time, can be modified to have an adjustable seat or foot stool so that the knees are higher than your hips, so as to protect the lower back.

2.4.2 Administrative controls

This involves changing the work practices and management policies, with the view of reducing prevailing workplace risks. Administrative control strategies are policies and practices that reduce WMSDs risk but they do not eliminate workplace hazards. Administrative controls are usually employed as a temporary measures until engineering controls can be implemented or when engineering controls are not technically feasible. Administrative improvements/controls include:

i. Changing work practices or the way work is organized.

ii. Providing variety in jobs - say by rotating workers through jobs that are physically tiring.

iii. Adjusting work schedules and work-pace - say by reducing shift length or limiting amount of overtime.

iv. Providing recovery time (i.e., muscle relaxation time);

v. Modifying work practices.

vi. Ensuring regular housekeeping and maintenance of work spaces, tools, and equipment.

vii. Regular health education seminar - for training in the recognition of ergonomic risk factors for WMSDs, and instructions in work practices and techniques that can ease the task demands or burden (e.g., stress and strain), avoiding static positions, awareness of proper lifting techniques.

viii. Changes in job rules and procedures such as scheduling more breaks to allow for adequate rest, and recovery.

2.4.3 Personal protective equipment (PPE)

Also known as Safety gear, PPE, generally provides a barrier between the worker and hazard source. Examples of PPE include: Respirators, ear plugs, safety goggles, chemical aprons, safety footwear (shoes), hard hats, knee and elbow pads. There are other devices (like braces, wrist splints, back belts, and similar devices), that are capable of reducing the duration, frequency, and/or intensity of exposure of risk factor for MSDs, although evidence of their effectiveness, as regarding offering personal protection against ergonomic hazards remains (i.e. injury reduction) inconclusive. In some instances, these devices may decrease one exposure, but increase another, because the worker has to "fight" the device to perform the work. An example is the use of wrist splints while engaging in work that requires wrist bending.

3. Musculoskeletal health

Soft tissues injuries (STI) of the musculoskeletal system are important cause of musculoskeletal ill health, particularly in working adult life. Thus, work may serve as a contributor to the development musculoskeletal ill health or exacerbator of an existing health condition. According to Health and Safety Executive (HSE),

WMSDs exert harmful effects on the life and well-being of workers in all fields, especially those requiring manual labor [24]. These injuries arise sequel to different causation:

i. *Distinct damage to tissues* - This is caused by instantaneous non-exertion events (i.e., intense physical exertion) like traumatic experiences (such as slips, trips, falls or whiplash following vehicle collisions;

ii. *Gradual damage to tissues* - This is caused sequel to exposure to occupational ergonomic stressors, such as repetitive and stereotyped motions, forceful exertions, awkward postures, vibration, and/or combinations of these exposures. These type of injuries are commonly referred to as WMSDs. Work-related musculoskeletal disorders have been severally reported to impair musculoskeletal health (MSH) due to the pain and lost of physical function while body tissues heal [25–27].

iii. *Insufficient interval of rest-time:* Sufficient interval of rest-time is generally needed for recovery, particularly between episodes of high usage; this will enhance the ability of the human body to repair itself. It is therefore advisable to incorporate sufficient interval of rest-time (as an ergonomic control) during task execution, in a work-place. Insufficient recovery time, combined with high repetitions, forceful movements and awkward postures, is capable of adversely altering MSH [26, 27].

Musculoskeletal health (MSH) means more than the absence of a musculoskeletal conditions. Good MSH implies that the musculoskeletal system (i.e., muscles, nerves, tendons, ligaments, joints, and cartilage) function well together without pain or discomfort. Thus, people with good MSH can carry out their functional activities (of daily living) with ease, and without discomfort. Poor MSH may be related to multiple risk factors like physical inactivity, being overweight or obese, diets deficient in vitamin D or calcium, smoking, older age and genetic predisposition to some musculoskeletal conditions [27]. Focusing on strategies directed towards reducing threats to MSH, sequel to work-place activities can therefore not be overemphasized; including early ergonomic evaluation of the work-place, to identify the risk (work-place/individual) factors; and early ergonomic intervention - directed at modifying physical work environments and work practices (**Figure 4**). This may be helpful in promoting MSH at the wok-place, and subsequently help prevent chronic pain, disability and work loss [28]. This is particularly important because there is a complex relationship between work and MSH. While it is true that certain types of work, and work-place conditions may have negative impacts on MSH, and may lead to failure to create a healthy environment and subsequently increase the risk of MSDs; undertaking meaningful work is an important part of an individual's sense of health and well-being [25]. Also, according to the National Institute for Health and Clinical Excellence (NIHCE), healthy workplaces provide an opportunity to promote good health generally, and musculoskeletal health specifically [26]. Furthermore, Waddell *et al.* [25] in their study titled "Work and common health problems", concluded that, overall, the beneficial effects of work outweigh the risks of work, and are greater than the harmful effects of long-term worklessness.

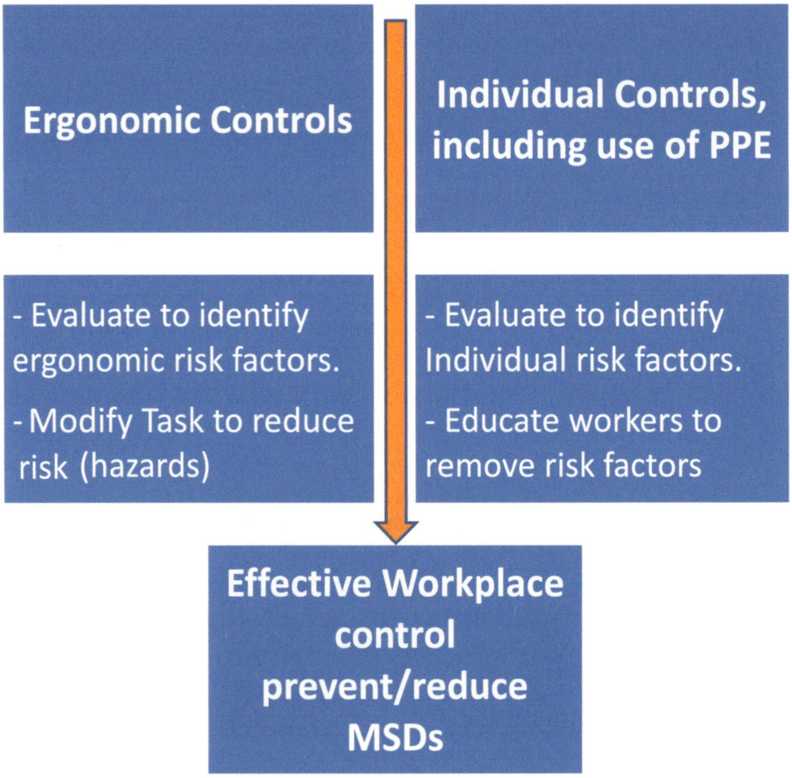

Figure 4.
Workplace ergonomic control of work-related musculoskeletal disorders.

4. Ergonomics principles

The definition of ergonomics (or human factors), as adopted by the International Ergonomics Association (IEA) in 2000 is "Ergonomics (or human factors) is the scientific discipline concerned with the understanding of the interactions among human and other elements of a system, and the profession that applies theory, principles, data and methods to design in order to optimize human well-being and overall system performance" [29]. Ergonomics is looked at from different perspectives; the word *"Ergonomics"* is derived from two words: *Ergon,* which means work and *Nomos,* which means laws. Ergonomics is also known as "human factors" and "human factors engineering". It supports the workers and their environment; vis-a-vis, the physical, psychological and social needs of the workers. Ergonomics is simply defined as the scientific study of human work (i.e., people at work). It is basically the science of fitting work-place conditions and job demands to the capability of a particular worker or working population, instead of fitting the workers to their work-place [1]. To achieve this, it important to consider the physical (and mental) capabilities, and limits of the worker as s/he interacts with the Tools/Equipment, Work methods, Tasks and Working environment. There are three major areas (or Dimensions) of ergonomics: Physical, Cognitive and Organizational (Environmental) aspects (i.e., dimensions) of ergonomics:

Physical Aspect of Ergonomics: Focuses on the physiological and bio-mechanical effects of work on human being e.g., Working postures, Work-stations, Work-related safety and health, Materials handled and Work-related musculoskeletal disorders.

Cognitive Aspect of Ergonomics: Focuses on the relationship between individual worker and the different systems (in the work-place) that workers (employees) operate with. It concerns the worker's cognitive processes, and the ability to process information, for instance technological solutions used at work.

Organizational Aspect of Ergonomics: Focuses on organizational processes, structures and policies at workplace, including communication within the workplace, working hours, work processes/methods and co-operation within operators.

4.1 Work-environment

Ergonomics is described in terms of the environment concerned, i.e. Work-Environments. Work is Physical or mental effort or activity directed towards the production or accomplishment of something (Task). It is the basis for skill acquisition; and is needed throughout all developmental stages for successful role function. While Environments include all the physical, chemical, and biological factors extended to human host, and all related behaviours, but excluding those natural environments that cannot reasonably be modified. Thus, Work-Environment describes: Circumstances, conditions, and influences that affect the behavior and performance of people/workers in the workplace. The following Physical factors affect job design - Noise, Vibration, Lighting, Temperature, Humidity, and Air Flow

4.2 Categories of work-environment

In terms of condition of job, there are two work-environments - *Office Work Environment (OWE)*, which specifically deals with the office environment; and *Industrial (Factory/Heavy-duty) Work Environment (IWE)*; which specifically deals with the Factory/Industrial environment.

While in terms of personality of the staff and/or management, there are also two categories: *Hostile Environment*, here, there appear to be little or no support what so ever, from the management and the direct boss. And *Friendly Environment*; here the workers received regular supports from the Management and their immediate boss. This may be created as an incentive for the workers.

4.3 Benefits of ergonomics

The primary goal of work-place ergonomics is to reduce workers' exposure to MSDs risk factors, thereby creating a safer and more healthful work environment. This is achieved through the use of effective WEP, which normally involves engineering and administrative controls. Work-place Ergonomics Principles (WEP) are directed towards designing workstations, tools and work tasks for safety, efficiency and comfort. This is ultimately, to prevent and/or control occurrence of MSDs, associated with the overuse of muscles (through force exertion), repeated tasks and assumption of awkward posture. Thus, effective WEP is directed towards preventing MSDs (injuries) by decreasing musculoskeletal imbalance and fatigue, associated workplace ergonomics risk factors like overuse of muscles (through force exertion), repeated tasks, and assumption of awkward posture; and failure to the workers body's recovery system, due to the assault of the ergonomic risk factors. This will invariably, result in increasing comfort (of doing work), job satisfaction and safety (of worker and Tools). Other benefits of workplace ergonomic include:

i. Increased productivity.

ii. Increased work quality.

iii. Increased efficiency at work.

iv. Reduced turnover (for both Machine & Human factor).

v. Reduced absenteeism.

vi. Reduced presenteeism

vii. Increased employee's morale.

viii. Decreased workers' compensation costs.

ix. Increased physical well being of the workers.

5. Work-place ergonomics

Work-place ergonomics involves designing the work-place, with consideration for the capabilities and limitations of the workers, with the view to reduce risk and promoting good musculoskeletal health; and consequently improve human performance and productivity, including office, recreation activities, and manual handling workplaces. According to General Duty Clause (GDC), Section 5(a)(1) of the Occupational Safety and Health Act (OSHA) regulation of 1970, employers have an obligation to keep the workplace free from recognized serious hazards (including ergonomic hazards), although OSHA regulations do not mandate an employer to provide ergonomic equipment such as work stations and chairs [30]. Common work-place risk factors (i.e., ergonomic hazard) include: Poor sitting posture, Awkward posture, Prolonged (stationary) positions, repetitive movements, Poor lifting (material handling), force/mechanical compression/vibration, temperature extremes, glare, inadequate lighting, and duration of exposure.

5.1 Ergonomic hazards at the office workplace

5.1.1 Poor (Sitting/standing posture)

Assumption of neutral postures when sitting (**Figure 5a** and **b**) and standing, is advised for ergonomic reasons, as poor postures exert uneven pressure on the spine, and the four natural curvatures of spine/backbone (i.e. cervical, thoracic, and lumbar spine) is preserved and maintained when lying, sitting, and standing. And may lead to uneven distribution of body weight. Therefore, efforts should be made to assume neutral postures, at all time; as this may be helpful in enhancing musculoskeletal health, and eventually help reduce the development of MSDs, like premature joint

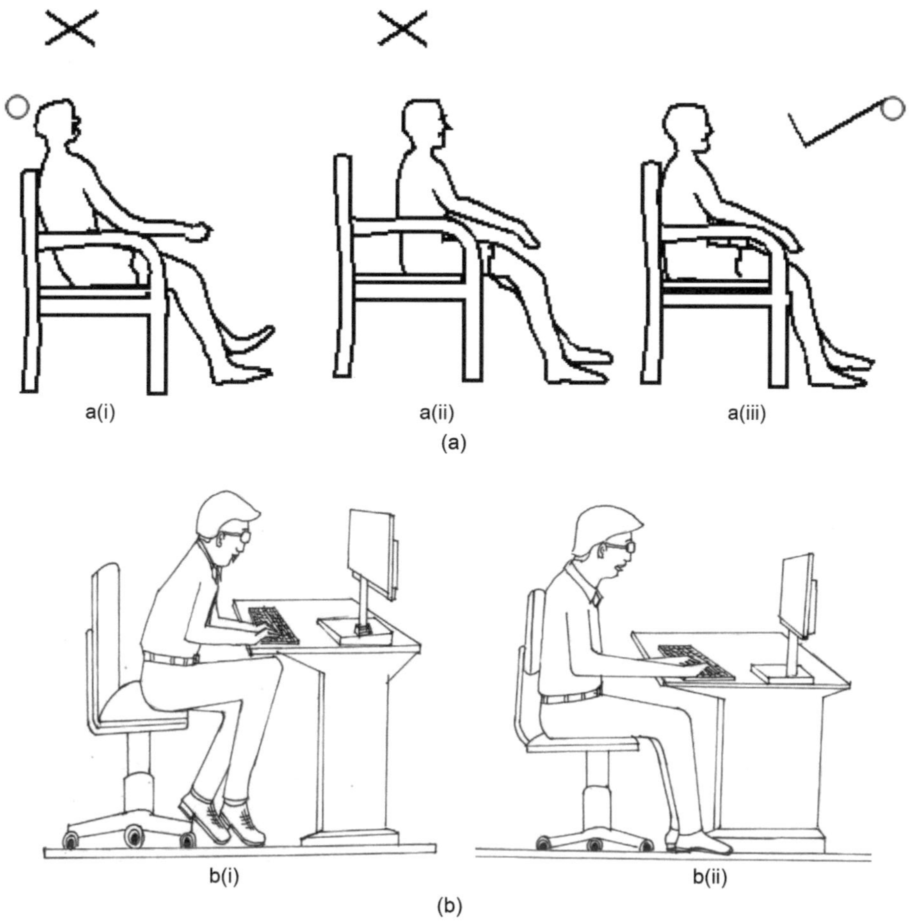

Figure 5.
(a) Three commonly assumed sitting postures a(i) – Ischio-femoral sitting posture, a(ii) – Sacra sitting posture, a(iii) – Sacro-femoral sitting posture [31]. b: Wrong b(i) and correct b(ii) office work-place sitting postures.

degeneration, nerve pinching and/or back pain. When assuming "*neutral sitting posture*", it is important to always be careful in ensuring that, the:

a. Back is straight with a slight inward curve (lordosis) of the low-back.

b. Neck and head are held upright, with the ears aligned with the shoulders.

c. Shoulders should be pulled back but relaxed.

d. Trunk (upper body) is not twisting or held leaning on one side.

e. Knees are bent at 90° and positioned slightly lower than the hips.

5.1.2 Awkward postures

These are the unnatural body positions (wrong postures) assumed at work (including poor manual handling), like bending, twisting, pocking of the neck - say, looking down at

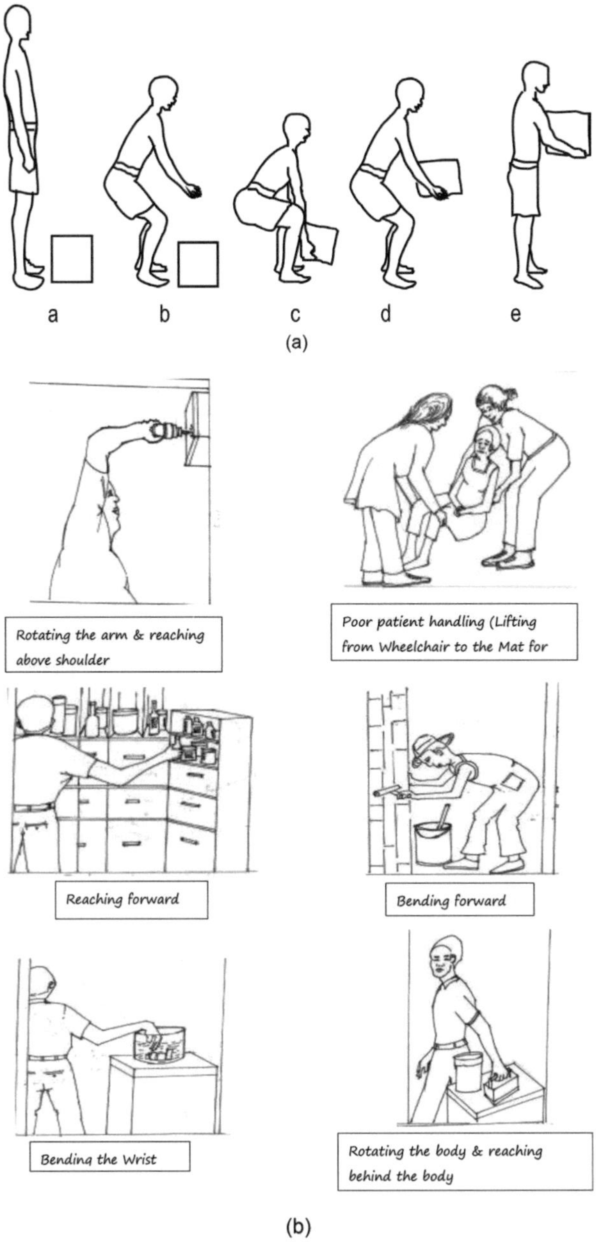

Figure 6.
(a) Illustration of proper lifting technique – a – starting position, b – bend your knees, keep your low back bowed-in, c – use correct grasp, d – use body weight to advantage, e – keep load close to the body [31]. (b) Some of the factors responsible for the development of MSDs.

your monitor, extending one's wrists to type, overreaching - say, to operate the computer mouse, and wrong bending - during manual handling (**Figure 6**). These postures push the affected joints past the mid-range of motion, thereby exposing the joints to ergonomic hazards, and subsequent injuries. Properly optimized workstation will minimize awkward postures, and this essentially involve assumption of good posture while at work. For optimal workstation, the following adjustment may be required:

a. Adjust the monitor height, such that the top line is at the level of the eye.

b. Adjust the chair height, such that the elbows are at an open angle (90 -110°) when typing.

c. Organize the workspace into zones, and keep frequently accessed items within arm's reach.

d. Set the arm-rests to desk level, this will further enhance better and easy placement of the wrists in neutral position.

e. Move/Swivel the chair instead of twisting the waist when rotating the body.

NB: When answering a phone call while at the work-station, the worker can use the speaker function instead of using the arm and/or shoulder wrongly, while holding up the phone.

5.1.3 Prolonged stationary position

This is prevalence among workers whose job tasks encourage sedentary lifestyle. Those who stay in the same position, for longer period of time. Usually sitting or standing in the same position for more than one hour is regarded as prolonged position [32]. Thus, prolonged stationary position is an ergonomic hazard; and it is advisable to observe regular breaks; of between 30 seconds to 5 minutes, usually every one hour, for stretching exercises at the work-place (**Figure 7**).

5.1.4 Frequent repetitive movements

Certain tasks require frequent repetitive movement. At the office space, the most common repetitive movements are performed by the fingers, wrist and arm. This can be observed in task involving the use of (operating) a mouse. The worker get to perform hundreds of small wind-shield movements with the wrist. This can be made worse, when the task is carried in an awkward position, and over a prolonged period of time. Performing repetitive motions repeatedly, however small, can cause micro-trauma to the surrounding tendons and tissues, consequently leading to the development of MSDs. Observing regular rest (particularly of the affected part); of between 30 seconds to 5 minutes, usually every one hour, for stretching exercises, in addition to eliminating the awkward posture is imperative. This allows the body to heal itself, making the worker to recover when the fatigue overturns workers body's recovery system. And will subsequently leads to the reduction in the development of MSDs.

5.1.5 Poor lighting at work

Insufficient lighting, unwanted dark spots and shadows, glare, and improper color temperature are some of the most common examples of poor lighting at work. They can negatively impact your vision, mood, and even productivity. It is important for workplace lighting to conform to good lighting ergonomics. Ensuring good lighting ergonomics includes the following:

i. The office should be arranged for natural light as much as possible.

Figure 7.
Therapeutic exercises at the workplace. (Each exercise is held or carried out for ten seconds, and repeated three times - for a complete bout of exercises).

ii. There should be adequate lighting (300–500 lux) around the immediate workspace.

iii. A combination of direct and indirect lighting should be used to eliminate shadows.

iv. The chair should be positioned, at a right angle from the windows to reduce glare.

v. Optimize your computer screen for good color and lighting contrast.

vi. Add diffusers to light fixtures to make them less harsh on the eyes.

vii. Monitor filters or computer glasses can be used to reduce blue light and glare.

6. Manual material handling (MMH), including patient handling

Manual material handling (MMH) is the process of routinely moving and handling of objects (including patients) through a series of biomechanical functions, such as; carrying, holding, lifting, pulling, pushing, and stooping on a regular basis. According to Manual Handling Operations Regulations (MHOR), MMH involve "any activity requiring the use of force exerted by a person to lift, push, pull, carry or otherwise move, hold or restrain an animate or inanimate object" [33]. By this definition, MMH does not excessively involve material handling only, it is also an integral to the practice of the physiotherapy, and rehabilitation professional generally, particularly in patient handling [33]. This is because the work-task of rehabilitation professionals (including physiotherapists) often requires the performance of physically demanding therapeutic activities, refers to as, patient handling tasks (PHT), that may constitute risk of developing MSDs. Patient handling tasks in rehabilitation are usually classified as "traditional" or "therapeutic:" Traditional PHT have a practical goal, like transferring a patient from bed to a wheelchair; "therapeutic" PHT, on the other hand, have more targeted goals, like facilitating patient function and independence. By and large, PHT have been widely reported to be capable of exposing rehabilitation professionals to high mechanical loads, particularly, on the spine [33–35].

Generally, manual handling frequently involve the performance of unsupported static posture, (which are often awkward in nature), during the execution of any particular work-task, including PHT; which usually involve postures like bending, reaching forward, twisting, squatting etc. (**Figure 6**). During the assumption of these awkward postures, the workers and/or worker's body parts are positioned away from their neutral position (**Figures 6**, **8** and **9**). Work-place ergonomic control is therefore essentially required; and directed towards maintaining a neutral body position, and to keep arms and legs as close to the trunk as possible (**Figures 8** and **9**). Also, modern patient handling technology is often recommended as part of a comprehensive safe patient handling programme in therapeutic and rehabilitation settings, in addition to the implementation of a standard and effective WEP [33, 36].

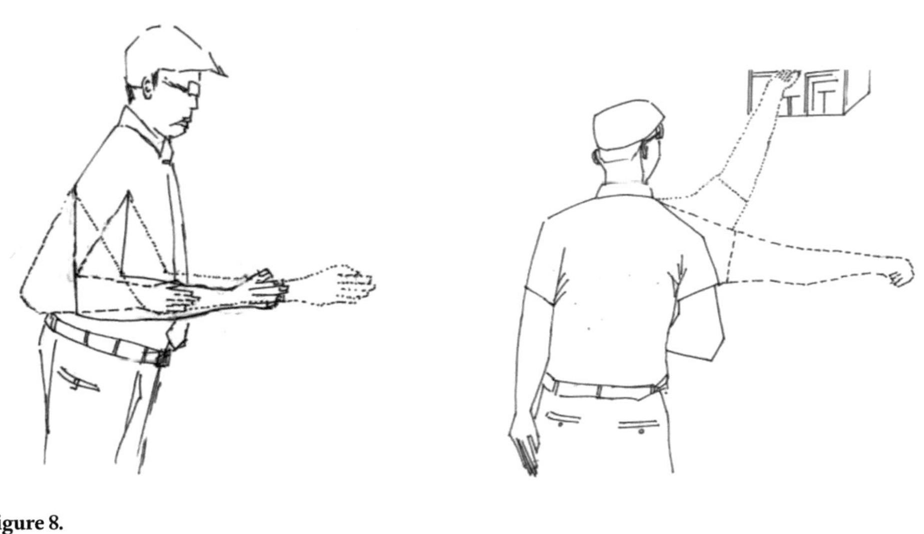

Figure 8.
Common awkward shoulder (Flexion/Extension, Abduction/Abduction/Extension) from neutral.

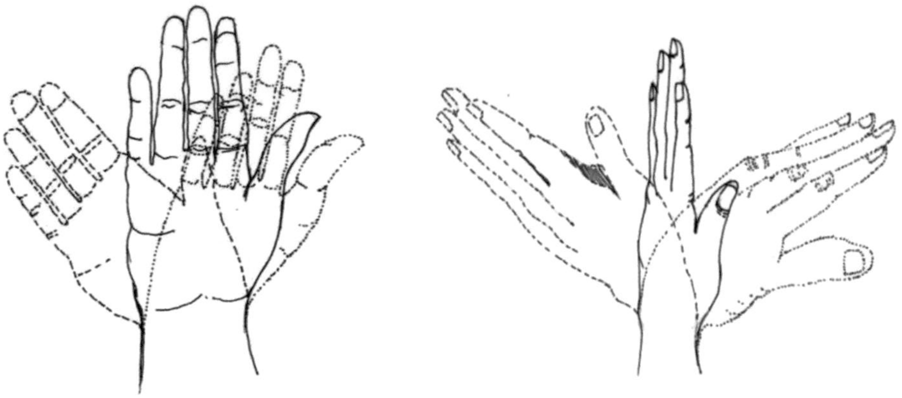

Figure 9.
Common awkward repetitive motions of the wrist (Radial/Ulnar deviation & Flexion/Extension, from the neutral).

6.1 Principles of safe manual handling

Basically, manual handling includes both transporting a load, and supporting a load in a static posture. The load may be moved or supported by the hands or any other part of the body, for example the shoulder. This may constitute ergonomic risks, particularly, if hazardous manual handling techniques are employed in the execution of the task (manual material and patient handling). In order to avoid work-place injury from hazardous manual handling, manual handling operations regulations require employers to:

a. *Avoid* hazardous manual handling, as much as possible;

b. *Assess* the risk of MSDs from any hazardous manual handling, particularly those that cannot be avoided;

c. *Reduce* the risk of MSDs from hazardous manual handling, where "reasonably practicable'; which is a measure of balancing the level of risk against the measures needed to control the risk in terms of money, time or trouble.

Furthermore, the risk of developing MSDs sequel to manual handling (materials and patients) can also be reduced by avoiding or reducing assumption of awkward postures (i.e., twisting, stooping, stretching etc.). This can be achieved by changing the:

1. *Task layout:* The best position to manually handle heavier loads is around waist height (**Figure 10**). Although, lighter loads, may easily be manually handled below knuckle height or above shoulder height, this should be practiced infrequently.

2. *Equipment used:* This include the use of mechanical assistance as handling aids. Here some manual handling is retained but bodily forces are applied more efficiently, reducing the risk of MSDs. Some examples; for material handling include Hand-powered hydraulic hoist, Roller conveyors, hydraulic lorry loading crane, Truck with powered lifting mechanism; and for patient handling, there are Patient standing hoist, Pulley and sling etc. [33, 36].

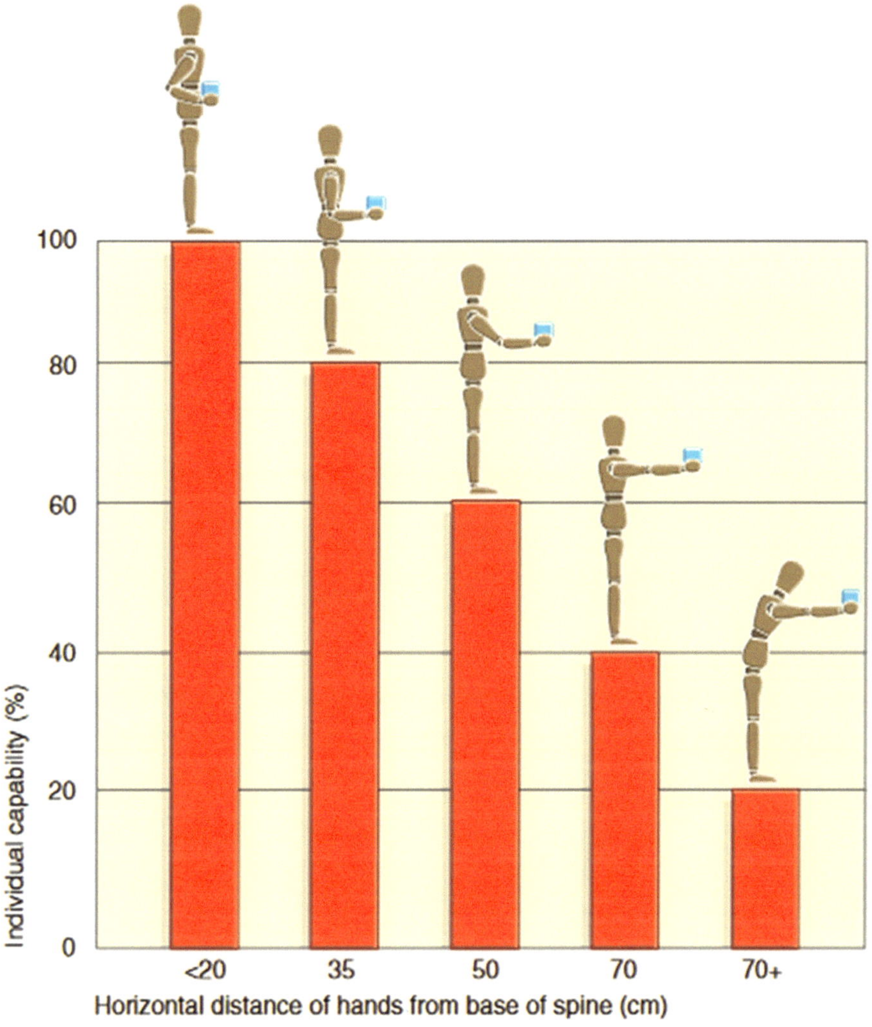

Figure 10.
Reduction of individual handling capability as the hands move away from the body [33].

3. *Sequence of operations:* by improving the flow of materials or products. There is a reduction in the individual handling capability, as the hands move away from the body (**Figure 10**).

It is imperative to note that the provision of a safe/good handling technique is no substitute for other risk-reduction steps, such as providing lifting aids, or improvements to the task, load or working environment. Example, moving a load (including patient) by rocking, pivoting, rolling or sliding is preferable to lifting it in situations where there is limited scope for risk reduction [33]. The principles of safe manual handling involve effective assessment of the task, and effective planning of execution of task, including, proper positioning of the body for effective handling, thus, ensuring proper positioning of the feet, securing a proper grip, and keeping the load close to the body, use the

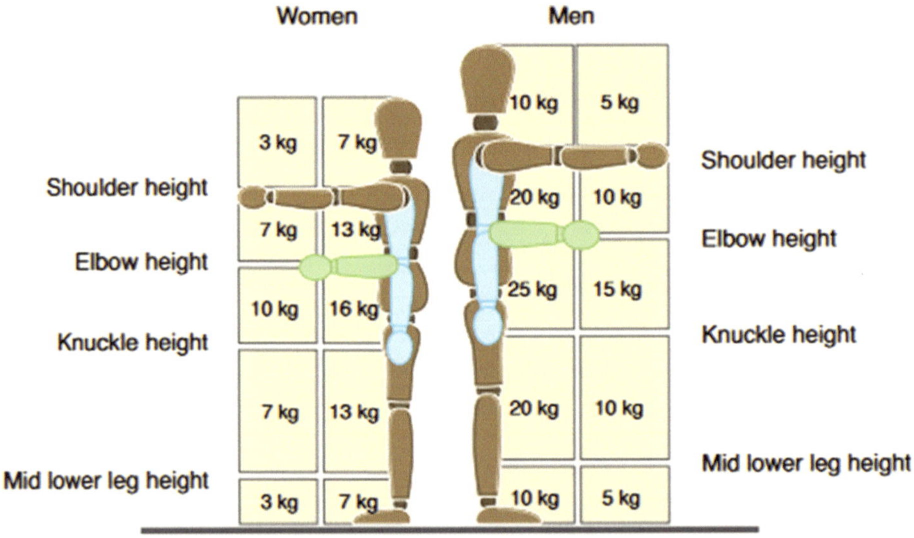

Figure 11.
Lifting and lowering risk filter – Each box contains a filter value for lifting and lowering in that zone. The filter values are reduced if handling is done with arms extended, or at high or low levels, as that is where injuries are most likely to happen [33].

leg muscles, with the view using body momentum to advantage (**Figures 10** and **11**). Assessment of manual handling risk is often anchored on four main areas, including the:

i. Nature of the task - including the workplace conditions, for example, the layout of the workstation and the speed of work (especially in conveyor-driven tasks/jobs.

ii. Load/Object being handled - Seize, type, weight, and any difficulty to grasp.

iii. Working environment the manual handling is taking place in: difficult to grasp, variations in level of floors or work surfaces, Space constraints, floors, temperature, ventilation and lighting

iv. Capabilities of the individual worker performing the manual handling. Individual handling capacity reduces as the hands move away from the body. Thus, as the load is moved away from the body, the level of stress on the lower back increases, regardless of the handling technique used (**Figure 10**). The workplace should be organized, such that the handler is as close to the load as possible. Extra caution should also be taken, when the employee concerns is new on the job, and/or has an underlining significant health problem or a recent injury. Manual handling is safe, when the lifter/handler hands

It is also important to access the weight of the load to be moved, and also to observe standard lifting technique (**Figure 10**). If it is less than the value given in the matching box, the operation is within the guidelines (**Figure 11**). If the lifter's hands enter more than one zone during the operation, use the smallest weight. If either the start or end positions of the hands are close to a boundary between two boxes you should use the average of the weights for the two boxes. According to [33]. The filter for lifting and lowering assumes: (a) the load is easy to grasp with both hands, (b) the operation takes place in reasonable working conditions, (c). the handler is in a stable body position.

NB: A good handling technique forms a very valuable addition to other risk-control measures. To be successful, good handling technique needs both training and practice. This may be helpful in reducing the risk of injury. The effects of these factors are interrelated, and may partly depend on the nature and circumstances of the manual handling operations. If the manual handling operations are carried out in circumstances which do not really change, like in manufacturing processes; the emphasis is particularly on improving the task and working environment. However, if the manual handling operations are carried out in circumstances which change continually, like certain activities carried out on construction sites, in delivery jobs, or in manual patient handling; the handler may offer less scope for improvement of the working environment and perhaps the task. Here, more attention is directed to the load - for example making the Load lighter for easier handling.

Some of the common workplace risk factors (i.e., ergonomic hazard) in Office work-place, are also considered in manual handling work-space, however, there are workplace ergonomics risk factors that are specifically of great consideration in MMH work-space. These include poor lifting techniques (material handling), wrong work-sitting, mix-match work-environment, and force/mechanical compression/vibration etc. (**Figures 5a, b** and **6a, b**)

7. Discussion and conclusion

This book chapter has provided a comprehensive description of Musculoskeletal system, the disorders of Musculoskeletal systems, causes and preventive strategies, using standard ergonomic principles, applicable at both materials manual handling work-place, and during patients management/rehabilitation. Ergonomic risk factors were discussed under two major categories, namely work-related (ergonomic) risk factors and individual-related risk factors. Work-related (ergonomic) risk factors were further discussed under two sub-categories - Primary and secondary risk factors. The primary work-related (ergonomic) risk factors are basically physical in nature, and they included Sudden/Forceful Exertions, High (Task) Repetitions, Awkward postures (Repetitive or Sustained), body vibration. The secondary risk factors include psychosocial factors (like work-pace, autonomy, monotony, work/rest cycle, task demands, social support from colleagues and management and job uncertainty). Other secondary risk factors include: Static Posture, Contact Stress, Cold/heat, Vibration, Noise, Physical Stress, Emotional Stress.

In order to evaluate the possibility of an employee developing WMSD, it is important to include all the relevant activities performed both at work and during leisure time activities (outside work). This is because, most of the WMSD risk factors can occur both at work and during leisure time activities. Also, because these ergonomic risk factors interact and act simultaneously, and has a synergistic effect on the musculoskeletal system, it is advisable and important to take into account this interaction rather than focus on a single risk factor.

Objectives

After studying this chapter readers should be able to:

- Describe the mechanism of the causes and prevention of WMSDs.

- Understand the different workplace environments, and the peculiar multiple risk factors (hazards) for WMSDs.
- Understand the importance of providing (for workers) an ergonomically sound work-environment, in the prevention WMSDs.
- Understand that WEP is more effective when ergonomic control is done at the workplace and during the performance of leisure time activities.
- Describe the advantages of considering the capabilities and limitations of the workers in the design of workplace.
- Identify the benefits of applying a three-tier hierarchy of controls (of Engineering, Administrative, and use of Personal Protective Equipment) as a standard intervention strategy against workplace hazards.
- Have a better understanding of principles of Safe Manual (Material and Patient) Handling.

Acknowledgements

The writing of this book chapter was influenced by the experience gathered from the teaching, particularly, of my Postgraduate students, at the department of Physiotherapy, college of Medicine, of the University of Lagos. Which started in 2009, when the Postgraduate Physiotherapy programme of the University of Lagos started. The author also appreciate Mr. Bamidele O. Olabisi of Biomedical Communication of College of Medicine, University of Lagos, Lagos, Nigeria, for the drawing of the illustrations

Conflict of interest

The author declare that this book chapter was written in the absence of any commercial or financial relationships that could be construed as a potential conflict of interest.

Notes/thanks/other declarations

N/A

DOI: http://dx.doi.org/10.5772/intechopen.106031

Author details

Daniel O. Odebiyi[1]* and Udoka Arinze Chris Okafor[2]

1 Faculty of Clinical Sciences, Department of Physiotherapy, College of Medicine, University of Lagos, Lagos, Nigeria

2 Department of Physiotherapy, College of Medicine, University of Lagos, Lagos, Nigeria

*Address all correspondence to: femiodebiyi@yahoo.com

References

[1] Bernard BP. U.S. Department of Health and Human Services, Centers for Disease control and Prevention, National Institute of Occupational Safety and Health. Musculoskeletal disorders and workplace factors: a critical review of epidemiologic evidence for work-related musculoskeletal disorders of the neck, upper extremity, and lower back. 1997. DHHS (NIOSH) Publication No. 97-141. Available from: https://www.cdc.gov/niosh/docs/97-141/

[2] Bureau of Labor: Statistics [BLS]. Nonfatal occupational injuries and illnesses requiring days away from work, 2014. Bureau of Labor: Statistics. U.S. Department of Labor, 2015

[3] Health and Safety Executive. Work-related musculoskeletal disorders statistics in Great Britain. Data up to March 2021. Annual statistics. Published 16th December 2021. Available at: www.hse.gov.uk/statistics/lfs/lfsilltyp.xlsx

[4] Adegoke BOA, Akodu AK, Oyeyemi AL. Work-related musculoskeletal disorders among Nigerian Physiotherapists. BMC Musculoskeletal Disorders. 2008;**9**:112

[5] Tella BA, Aiyejusunle CB, Adekunle KM, Odebiyi DO, Okafor UAC. Work related musculoskeletal disorders among physiotherapists in Nigeria. Ghana Journal of Physiotherapy. 2009;**1**(1):6-9

[6] Tinubu BMS, Mbada CE, Oyeyemi AL, Fabunmi AA. Work-related musculoskeletal disorders among nurses in Ibadan, South-west Nigeria: A cross-sectional survey. BMC Musculoskeletal Disorders. 2010;**11**:12

[7] Odebiyi DO, Akanle OT, Akinbo SR, Balogun SA. Prevalence and impact of work-related musculoskeletal disorders on job performance of call center operators in Nigeria. The International Journal of Occupational and Environmental Medicine. 2016;7(2):98-106

[8] Kaka B, Idowu OA, Fawole HO, Adeniyi AF, Ogwumike OO, Toryila MT. An analysis of work-related musculoskeletal disorders among butchers in Kano Metropolis, Nigeria. Safety and Health at Work. 2016;7(3):218-224

[9] Omojunikanbi OA, Akinpelu AO, Ekechukwu END. Prevalence, pattern and predictors of work-related musculoskeletal disorders among oil workers in Nigeria. Work. 2022;**71**(1):151-163

[10] NIOSH. Musculoskeletal disorders and workplace factors. A critical review of epidemiologic evidence for work-related musculoskeletal disorders of the neck, upper extremity, and low back. NIOSH 1997. Publication no. 97-141

[11] Pavlovic-Veselinovic S, Hedge A, Grozdanovic M. An expert system for risk assessment of work-related musculo-skeletal disorders. In: Khalid H, Hedge A, Ahram TZ, editors. Advances in Ergonomics Modeling and Usability Evaluation. Boca Raton: CRC Press; 2010

[12] Bugajska J, Sagan A. Chronic musculoskeletal disorders as risk factors for reduced work ability in younger and ageing workers. International Journal of Occupational Safety and Ergonomics. 2014;**20**(4):607-615. DOI: 10.1080/10803548.2014.11077069

[13] Rufa'i AA, Oyeyemi AL, Maduagwu SM, Fredrick AD,

Saidu IA, Aliyu SU, et al. Work-related musculoskeletal disorders among Nigerian police force. Nigerian Journal of Basic Clinical Science. 2019;**16**:127-133

[14] Skovlund PC, Nielsen BK, Thaysen HV, et al. The impact of patient involvement in research: A case study of the planning, conduct and dissemination of a clinical, controlled trial. Research Involvement and Engagement. 2020;**6**

[15] NIOSH workers health chartbook. NIOSH Publication No. 2004-146. Washington, D.C. 2004

[16] McCauley-Bush P. Ergonomics: Foundational Principles, Applications and Technologies, an Ergonomics Textbook. Boca Raton, FL: CRC Press, Taylor & Francis; 2011

[17] World Health Organization [WHO]. Identification and control of work related diseases. Technical report series n° 714, World Health Organization, 1985

[18] Sauter S, Hales T, Bernard B, Fine L, Petersen M, Putz-Anderson V, et al. In: Luczak H, Cakir A, Cakir G, editors. Summary of two NIOSH field studies of musculoskeletal disorders and VDT work among telecommunications and newspaper workers. Elsevier Science Publishers; 1993

[19] Punnett L, Wegman DH. Work-related musculoskeletal disorders: The epidemiologic evidence and the debate. Journal of Electromyography and Kinesiology. 2004;**14**(1):13-23

[20] Okunribido O, Wynn T. Ageing and work-related musculoskeletal disorders. A review of the recent literature, HSE, 2010

[21] Buckle P, Devereux J. Work Related Neck and Upper Limb Musculoskeletal Disorders. Bilbao, Spain: European Agency for Safety and Health at Work; 1999

[22] Larsson SE, Bodegard L, Henriksson KG, Oberg PA. Chronic trapezius myalgia: Morphology and blood flow studied in 17 patients. Acta Orthopaedica Scandinavica. 1990;**61**(5):394-398

[23] Putz-Anderson V. Cumulative Trauma Disorders: A Manual for Musculoskeletal Disorders of the Upper Limbs. Philadelphia: Taylor & Francis; 1988

[24] Health and Safety Executive. Work-related Stress, Depression or Anxiety Statistics in Great Britain., 1-11. [Internet] 2017. Available from: www.hse.gov.uk/statistics/ [Accessed May 26, 2022]

[25] Waddell G, Aylward M, Sawney P. Work and common health problems. Journal of Surgical Medicine. 2007;**39**(2):109-120

[26] National Institute for Health and Clinical Excellence. Promoting physical activity in the workplace [PH13], 2008

[27] Booth FW, Roberts CK, Laye MJ. Lack of exercise is a major cause of chronic diseases. Comprehensive Physiology. 2012;**2**(2):1143-1211. DOI: 10.1002/cphy.c110025

[28] Linton SJ. Early identification and intervention in the prevention of musculoskeletal pain. American Journal of Industrial Medicine. 2002;**41**(5):433-442

[29] International Ergonomics Association [IEA]. Definition of Ergonomics, 2000. [Internet] Available from: https://iea.cc/what-is-ergonomics/ [Accessed June 02, 2022]

[30] Abrams AL. OSHA's General Duty Clause: A Guide to Enforcement and Legal Defenses. In: Paper presented at the ASSE Professional Development Conference and Exposition, Dallas, Texas. 2015

[31] Odebiyi DO. How to Avoid Back Pain at Work and at Rest: A Handbook of the Nigerian Back School [NBS]. Crowntex Printing Press; 2004. ISBN: 978-059-548-1

[32] McKenzie RA. The Lumbar Spine: Mechanical Diagnosis and Therapy. New Zealand: Spinal Publication Walkanae; 1981

[33] Manual handling at work: A brief guide Leaflet INDG143 (rev3) HSE [Internet] 2012. Available from: www.hse.gov.uk/pubns/indg143.htm. [Accessed: June 02, 2022]

[34] Waters TR, Rockefeller K. Safe patient handling for rehabilitation professionals. Rehabilitation Nursing: The Official Journal of the Association of Rehabilitation Nurses. 2010;**35**(5):216-222

[35] Darragh AR, Shiyko M, Margulis H, Campo M. Effects of a safe patient handling and mobility program on patient self-care outcomes. The American Journal of Occupational Therapy. 2014;**68**(5):589-596

[36] Ajulo OO. Evaluation of working postures of physiotherapists in Lagos state. An MSc Dissertation, to the Department of Physiotherapy, School of Postgraduate Studies, University of Lagos, in partial fulfillment of the award of Masters of Physiotherapy (Ergonomic Specialty), 2010

Chapter 3

Musculoskeletal Disorders in the Teaching Profession

Patience Erick, Tshephang Tumoyagae and Tiny Masupe

Abstract

Musculoskeletal disorders (MSDs) are among the most common and important occupational health problems in working populations with significant impact on quality of life and a major economic burden from compensation costs and lost income. MSDs decrease productivity at work due to absenteeism, presenteeism and sick leave. During the course of their work, teachers can be subjected to conditions that cause physical and psychosocial illness. Common MSDs among teachers include those affecting the lower back, neck and upper extremities. Research suggests that the aetiology of MSDs is complex and multifactorial in nature. Occupational factors including location of school, carrying heavy loads, prolonged computer use, awkward posture and psychosocial factors such as poor social work environment, high anxiety and low job satisfaction have been found to contribute to development of MSDs. Factors such as high supervisor support and regular physical exercise on the other hand have been found to have a protective effect against MSDs among teachers. The interventions for these conditions need to be contextualized for them to be effective and to take into consideration, the risk factors for these conditions and how they interact with each other.

Keywords: musculoskeletal disorders, low back pain, neck pain, lower limb pain, teachers

1. Introduction

Musculoskeletal disorders (MSDs) represent one of the most common and most costly occupational health problems globally [1]. Developing countries are disproportionately affected where working conditions could be poor due to acute lack of awareness on ergonomic issues, education and training [2]. MSDs have also been associated with high levels of health-related presenteeism, absenteeism and sick leave among teachers [3]. MSDs are conditions that affect the body's muscles, joints, tendons, ligaments, nerves, bones and their local blood supply. Most work-related MSDs develop over time and caused by either work itself or the worker's working environment [4].

School teachers, in general, have been shown to report a high prevalence of MSDs relative to other occupational groups [5], with prevalence rates of between 40% and 95% according to a systematic review conducted in 2011 [1]. These high prevalence rates among teachers are associated with individual, work-related and psychosocial factors. Some studies have investigated the relationship between MSDs in teachers and their working conditions. The work tasks of teachers involve a wide variety of duties and responsibilities that may involve prolonged sitting and standing, use of inappropriate furniture, awkward postures likely adopted when writing on the board, helping students with their work or when helping students during extracurricular sporting activities. Furthermore, teachers might adopt awkward postures when reading, marking students' work or preparing lessons. The constant loading of the muscles in the neck, shoulders and the back eventually leads to aches, pains or discomfort [5, 6]. In some instances these activities may be carried out under unfavorable working conditions. Psychosocial risk factors such as poor mental health, low supervisor or colleague support, low job satisfaction, high job stress and high psychological job demands have also been associated with development of MSDs [1]. Preventive programmes are required for management of these disorders and this should ideally be at organizational level rather than individual level [7].

2. Assessment of MSDs among teachers

Historically, evaluation of MSDs has involved use of many different methods ranging from broad approaches to specific techniques. Widely accepted approaches for determining the prevalence of MSD and favoured by researchers include self-developed questionnaires [8–10] and the Standardized Nordic Questionnaire [11–14]. Self-developed questionnaires can be structured, semi-structured or unstructured [8] and employ open vs. closed, single vs. multiple responses, ranking, and rating [15, 16]. The Standardized Nordic questionnaire was developed by a Swiss company for analysis of musculoskeletal symptoms. It has both the General the Specific Questionnaire. The General Questionnaire is a graphic in which the human body is split into nine anatomical regions, whereas the Specific Questionnaire focuses on anatomical locations where musculoskeletal problems are more common [9]. The fundamental benefit of using these questionnaires is that they examine the severity of symptoms, their impact on work and leisure activities, the overall duration of symptoms, and sick leave.

Other MSDs evaluation methods include pilot study surveys and questionnaires like the pilot tested surveys, [17–19], questionnaires such as the Northwick Neck Pain Questionnaire [20], Health Questionnaires [21], Job Content questionnaires [22] and the Subjective Health Complaints Questionnaire [6]. While questionnaires are a cost-effective and manageable method to collect data, they can create recall bias and make follow-up difficult, particularly when anonymous reporting is used [1]. Disregard of physical examination and assessment pervades diagnosis of MSD even though they could likely produce more accurate results. These methods are considered expensive and time consuming and therefore rarely used [1].

The majority of the research used self-reported questionnaires to assess MSDs. Self-reporting has limitations such as participants not being honest, introspective inability, wrong interpretation of questions, recall [23] and sampling bias. The participant recall bias could lead to under or overestimation of MSDs [24]. Additionally, self-reporting could lead to respondents reporting all pain as MSDs [25]. It is also not possible to establish any causal-effect associations through self-reporting [26]. The presence of MSDs is only dependent on the participants' self-reports and not on an objectively validated diagnosis.

3. Prevalence of MSDs among teachers

Following the systematic review on MSDs among school teachers that was done in 2011 by Erick and Smith [1], substantive research has been carried out on the subject. Globally studies have been conducted among nursery to secondary school teachers on MSDs generally and/or on specific body sites such as neck and/or shoulder, back, upper and/or lower limbs. Recently it was estimated that approximately 1.71 billion people globally have musculoskeletal conditions [27]. A previous systematic review of MSDs among school teachers which was based on papers published between 1981 and 2011 revealed that these conditions affect between 39% and 95% of teachers [1]. The prevalence rates of MSDs among school teachers reported on studies carried out after this review range between 21.1% and 96%.

3.1 Global prevalence of MSDs among teachers

3.1.1 Prevalence of MSDs in Asia

A substantive amount of research on MSDs among teachers has been conducted in the past ten years in Asia. The prevalence rates of general MSDs in the region range between 21.1% and 93.7% with high prevalence rates of 90.7–93.7% reported among school teachers in China [28]. Similarly, 87.3% and 80.1% of secondary school teachers in Saudi Arabia [23] and primary school teachers in Malaysia, respectively [29] reported ever experiencing MSDs. Furthermore, in a study carried out in Pakistan [30] and another study from Saudi Arabia [31], 82.7% and 79.2% female school teachers reported MSDs, respectively. Prevalence rates of MSDs ranging between 60.3% and 74.5% were reported among school teachers in other Asian countries [8, 32–35]. Low prevalence rates of MSDs have been reported among primary teachers in another study conducted in Malaysia (40.1%) [36] and male secondary teachers in Saudi Arabia [21].

3.1.2 Prevalence of MSDs in South America

Relatively few studies have been carried out to investigate the prevalence of MSDs among teachers South America. In Chile, the 12-months prevalence of MSDs among school teachers was 88.9% [26] while in Bolivia it was 86% [37]. A 7-days MSDs prevalence of 63% was reported among Bolivian school teachers [37]. In a study of chronic musculoskeletal pain among Brazilian teachers in Londrina, 43% reported experiencing chronic pain in the past 12 months [38].

3.1.3 Prevalence of MSDs in Africa

A high prevalence rate of MSDs was reported among Egyptian nursery school teachers (96%) [39]. In Botswana, a 12-months prevalence of MSDs among primary and secondary school teachers was 83.3% [10]. The prevalence rates of MSDs among teachers in two studies carried out in Ogun State [10] and Enugu State, Nigeria [41] were 70.47% and 70.2% [41], respectively.

As reflected above, MSDs appear to be highly prevalent in the teaching profession with the high prevalence reported among nursery schools. MSDs studies carried out in Europe were specific to different body sites.

3.2 Prevalence of MSDs according to different body sites

3.2.1 Neck and/or shoulder pain

Although most of the studies investigated 'neck pain' and 'shoulder pain' separately, some combined these and reported on them as neck and/or shoulder pain (NSP). In a study conducted in Durban, South Africa a 12-months prevalence of NSP among primary school teachers was 80.4% [42], In Chile, 68.6% of school teachers reported NSP in the last 12 months [26]. Similarly, in two separate studies conducted in Malaysia, 60.1% of secondary [43] and 56.5% of primary [29] school teachers reported NSP. Parallels could be drawn to a study conducted in Ethiopia where 57.3% of teachers reported NSP [13]. In a study conducted in China, almost half (48.7%) of school teachers reported experiencing NSP in the previous 12 months [44]. These studies show that MSDs of the neck/shoulder are highly prevalent.

3.2.2 Neck pain

Even when neck pain is reported separately, there is still evidence that it is a prevalent MSD among school teachers with studies from different countries reporting prevalence rates above 50% with high levels reported in Turin, Italy at 75.6% [45]; followed by Nigeria at 57% [41, 46], Botswana at 50.2% [40] and Bolivia at 47% [37]. Other countries however reported low prevalence rates of neck pain among teachers. Low prevalence of neck pain have been reported among Saudi female teachers (11.3%) [47] and Nigerian teachers (3.2%) [10].

3.2.3 Shoulder pain

High prevalence rates of shoulder pain were reported by teachers in China (73.4%) [44]. Parallels could be drawn to the results of a study that was conducted in Nigeria where 62.3% of teachers reported shoulder pain. Most of the studies reported prevalence rates ranging between 41% and 57.5% [12, 21, 23, 30, 32–34, 46, 48, 49]. However, low prevalence rates were reported in studies conducted among female teachers in Malaysia (22.2%) [36] and Saudi Arabia (20.6%) [47] and primary school teachers in Egypt (15%) [50] and Ogun State, Nigeria (11.7%) [10]. Lessons could be learnt from these places on factors associated with these low rates of shoulder pain.

3.2.4 MSDs in the upper extremities

When compared to other MSDs, upper extremities appear to be less prevalent. A study that was conducted in Brazil reported that 14% of teachers experienced upper limb pain [38]. Wrist/hand pain was reported by 26% teachers in Chuquisaca, Bolivia [37], 23.4% in Turkey [51] and 16.2% secondary female teachers in Saudi Arabia over 6 months [31] and 7.4% in another study conducted among female teachers in Saudi Arabia over 3 months [47]. Elbow pain on the other hand was reported by between 5.6% and 16% of teachers in studies carried out in Italy [45], Turkey [51], and Malaysia [36]. Although prevalence rates for MSDs of the upper extremities were generally low across most countries, there were a few countries where prevalence rates could go above 40% as was the case among primary school teachers in Kota Kinabalu, Malaysia who reported hand/fingers pain in the last 6 months [29].

3.2.5 Low back pain

In this section, the prevalence of back pain among school teachers is discussed. Limited studies reported general back pain whilst majority separated low back pain and upper back pain. The prevalence of general back pain was reported in studies conducted in Qassim, Saudi Arabia (74.4%) [11], Minas Gerais, Brazil (58%) [52], Turkey (42.7%) [34] and Iran (39%) [12].

When compared to other MSDs among school teachers, low back pain (LBP) appears to have been the most studied. High prevalence rates of low back pain have been reported in studies conducted in Spain [53], Jordan [54] and Ekpoma State, Nigeria [46] where 96.5%, 92.3% and 85% school teachers reported low back pain respectively. Almost three-quarters of teachers in Turkey (74.9%) [34], Northern Ethiopia (74.8%) [55], Putrajaya Malaysia (72.9%) [9] and Italy (70.6%) [45] reported experiencing LBP in the past 12 months. Most of the studies reported prevalence rates between 35.3% and 68% [14, 23, 24, 31, 32, 42, 47, 56, 57]. However, some studies reported low prevalence rates. One quarter of female teachers in Terengganu, Malaysia [36] and school teachers in Kanpur, India [58] reported LBP in the previous 12 months. In Abha City, Saudi Arabia [21] and Ogun State, Nigeria [10], one fifth of teachers reported LBP while in Brazil 13% of them also reported LBP [38]. LBP is common among teachers regardless of the geographical location. This is a concern as LBP is a leading cause of disability in both developing and developed countries [44].

3.2.6 Upper back pain

Upper back pain does not appear to have been studied as much as LBP. Although 84% of preschool teachers in Turin, Italy reported experiencing upper back pain [45] it appears this pain is not as prevalent as LBP. This is evidenced by prevalence rates reported in studies carried out in Enugu State in Nigeria [41], Peshawar, Pakistan [30] and Thailand [33] where 47.4%, 43.3% and 36.1% of school teachers reported upper back pain, respectively. In Terngganu, Malaysia, one quarter of female primary school teachers reported upper back pain experienced in the previous 12 months [36]. Lower prevalence rates were reported among female Saudi teacher (17.7%) [47] and teachers in Ogun State in Nigeria (1.1%) [10].

3.2.7 MSDs of the lower extremities

Several studies have investigated MSD in the lower extremities such as the knees, leg, hips, ankles and feet. In a study conducted in Kota Kinabalu, Malaysia, almost half of the primary school teachers reported lower extremities pain in a period of 6 month [29]. However, a lower prevalence of 13% was reported in a study of Brazilian teachers [38]. The prevalence rates of knee pain among different school levels ranged between 26.3% and 49%. About 49% of nursery school teachers in Ekpoma State in Nigeria reported knee pain [46] while it was reported by 41% of secondary school teachers in Hail, Saudi Arabia [23]. Parallels could be drawn to the results of studies conducted in Enugu State of Nigeria [41] and Turin, Italy [45] where 39.3% and 38.7% of teachers and nursery school teachers reported knee pain respectively. One third of teachers in Turkey [34] and Terengganu, Malaysia [36] reported knee pain in the last 12 months while in Saudi Arabia one quarter reported the same condition in the past 3 months [47].

The prevalence rates of leg pain among teachers ranged between 38.7% and 65.2%. The highest prevalence rate was reported by school teachers in Ogun State in Nigeria [10] while the lowest was reported by preschool teachers in Turin, Italy [45]. Hip pain was reported by between 15.4% and 45.3% of teachers in Enugu State in Nigeria [41], nursery schools in Ekpoma State in Nigeria [46], female teachers in Saudi Arabia [47] and teachers in Turkey [34]. Some studies combined hip and thigh pain and was reported by 49.6% preschool teachers in Italy [45] and 18.4% of teachers in Terenggamu, Malaysia [36] in the past 12 months.

The prevalence of ankle pain was relatively common among teachers ranging between 12.3% and 48.4%. Female teachers in Pakistan reported the highest prevalence rate of this pain (48.4%) [30]. Although nursery school teachers have been thought to be at increased risk of ankle pain due to activities which require sustained periods of kneeling, stooping, squatting or bending [59], only 31% of nursery school teachers in Ekpoma State, Nigeria reported ankle pain [46]. Some studies studied ankle and feet pain combined and the highest prevalence (85.5%) was reported in a study conducted among school teachers in Abha City, Saudi Arabia [60]. However, relatively low prevalence was reported in studies carried out in Terenggamu, Malaysia [36] and preschool teachers in Italy [45] where this pain was reported by 32.5% and 16.8% of the study population, respectively.

MSDs have been previously reported to be more prevalent among nursery school teachers because of the kind of work they do. This chapter confirms the previous findings because when compared to other school teachers, high numbers of nursery teachers reported general MSDs, upper back pain, neck and/or shoulder, knee and elbow pain. This has been attributed to that nursery school teachers perform a wide variety of tasks and combine basic health childcare and teaching duties, and those that require sustained mechanical load and constant trunk flexion [59, 61]. Furthermore, nursery school teachers have been found to have elevated prevalence of MSDs due to activities which require sustained periods of kneeling, stooping, squatting or bending [59]. The high prevalence of MSDs of different body sites among teachers is a concern as this population consists of high numbers of members of the society. Teachers ill-health does not only affect them but high likely to affect learners. Therefore, it is crucial to establish work-related factors that affect this population to put in place control measures that will reduce prevalence and progression of these conditions. The following section discusses work-related factors associated with teachers reporting MSDs.

4. Work-related risk factors for MSDs among teachers

A large proportion of MSDs have been associated with adverse work conditions. Increased risk of these disorders have been reported in occupations with repetitive work tasks, awkward postures and heavy lifting as well as psychologically demand work environments. The section will discuss work-related factors associated with MSDs among school teachers.

4.1 Location of school

In a study carried out in both rural and urban areas of Bolivia, teachers working in rural areas were more than two to almost four times more likely to report any work limiting musculoskeletal pain during the last 12-months and for work limiting pain in at least three parts of the body than teachers in urban areas. The study also found that work limiting pain in ankles was higher in rural than urban school teachers [37]. These findings have been attributed to that apart from the teaching responsibilities, teachers in rural areas work closely with the rural communities which could be both physically and psychologically demanding. For this reasons, there is often concentration of professionals in urban areas as opposed to rural areas which in turn impacts on the quality of education and increased inequalities between the two areas [37].

4.2 Carrying heavy loads

Carrying heavy loads have also been associated with MSDs among school teachers. Brazilian teachers in Londriana region who reported carrying didactic materials were almost two times more likely to report upper limbs pain than those who did not report carrying heavy materials [38]. Lifting loads with hands was also associated with LBP among secondary school teachers in Putrajaya, Malaysia. Teachers who reported lifting loads with hands were at increased risk of developing LBP than those who did not report so [9]. Carrying weight has also been significantly associated with MSDs among secondary school teachers in Fiji [62].

4.3 Prolonged computer use

Prolonged computer use has previously been associated with MSDs of different body sites among school teachers. Brazilian teachers in Minas Gerais region who reported using computer or tablet within 5 h and for 6 or more hours during the COVID -19 pandemic were 1.12 times and 1.27 times more likely to report back pain compared to those who did not report computer or tablet use [52]. Primary school teachers in Samsun Turkey who reported daily computer use were at increased risk of neck pain when compared to those who did not indicate daily computer use [34]. Prolonged computer use leads to prolonged sitting. Activities of prolonged sitting and computer use are unsafe acts favorable for the development of neck/shoulder pain, back pain and upper limb pain among teachers [44]. This may also be attributed to a sustained forward head posture and/or constant neck flexion which cause static overload of neck and shoulders muscles. When combined with repetitive movements associated with a mouse, touchpad or keyboard can increase the likelihood of shoulder and/or neck pain [48].

4.4 Awkward postures

Awkward postures have been found to contribute to reporting of MSDs. This is evidenced by the results of a study among primary school teachers in Cairo, Egypt where awkward posture was associated with MSDs [50]. Furthermore, teachers who reported awkward arm position at work in a study conducted in Botswana were 1.4 times more likely to report LBP than those who did not report awkward arm position [63]. In another study conducted in Botswana, teachers who had reported awkward arm position when working were at risk of shoulder pain, upper back pain and wrists/hands pain [40]. Teachers in Gondar town of Ethiopia who had reported static head down posture and elevated arm over shoulder were 2.26 times and 2.71 more likely to report shoulder/neck pain than those who did not report the awkward postures [13]. Similarly, Chinese teachers who reported prolonged static posture were more likely to develop NSP and LBP than those who did not report static posture. Teachers who reported that they acquired posture characterized by twisting were also at increased risk of LBP than those who did not report so [44]. Bending has been significantly associated with MSDs among secondary school teachers in Fiji [62].

Stretching to write on the board placed school teachers in Thailand at increased risk of repetitive strain injuries [33]. Writing on the board has also been strongly associated with MSDs of different anatomical areas such as upper limbs pain, LBP and lower limbs pain among Brazilian teachers in Londrina region [38]. Forward-bending and backward bending of the head for a prolonged time when writing on the board has been significantly associated with NSP among primary school teachers in South Africa [42]. Shoulder pain may occur as a result of working with raised arms unsupported for a considerable time, a characteristic synonymous with teachers' work as they write on the board. Awkward postures caused by sustained muscle stretching particularly overhead are likely to induce neck and/or upper limbs pain in teachers. Awkward postures affect MSDs of different body areas. This is so because the broad activities which teachers participate in such as reading, marking, lesson preparation lead to prolonged sitting periods, bending to assist students at student level, writing on and reading from the board put strain on different body areas.

Prolonged sitting and standing have been associated with MSDs of different body area. A study of Chinese teachers in Guang dong Province, found that those who reported prolonged sitting were at risk of reporting NSP and LBP than those who did not report prolonged sitting. NSP was also experienced by those teachers who reported prolonged standing than those who did not [44]. Prolonged standing has also been associated with LBP among Egyptian teachers [55] and general MSDs among male secondary school teachers in Saudi Arabia [32]. Prolonged sitting has also been significantly associated with NSP among Gondar teachers in Ethiopia [13], with LBP among secondary school teachers in Putrajaya, Malaysia [9] and among primary school teachers in Durban, South Africa [42] and foot pain among Saudi teachers in Abha Sector [60]. Prolonged standing and sitting were also significantly associated with MSDs among female school teachers in Pakistan [30] and secondary school teachers in Fiji [62]. Standing and sitting for a long period, working in a head down posture for long periods, bending/twisting upper body have been significantly associated with MSDs among preparatory government school teachers in Cairo, Egypt. The study further found that prolonged working in the same posture, helping students into flexing posture and repeating the same movement of arms or hands many times per minute were also significantly associated with MSDs [39].

4.5 Inappropriate furniture

Previous research indicates that inappropriate furniture contributes to development of MSDs. A significant association has been found between MSDs and school furniture among school teachers in two separate studies carried out in Egypt [39, 50]. Uncomfortable work chair/table was significantly associated with MSDs among female school teachers in Pakistan [30]. In China, school teachers who reported uncomfortable back support were about two times more likely to report NSP and LBP compared to those who did not report so [44]. Women teachers, nurses and sonographers in Sweden who reported that they were dissatisfied with computer workstation arrangements were 1.2 times more likely to report neck pain and shoulder pain respectively than those who reported that they were satisfied [64].

Similarly, school teachers in Enugu State of Nigeria who reported using teaching board with height of 180–190 cm and more than 190 cm were 3.5 times and 4.6 times more likely to report neck pain, respectively than those who used teaching board that was less than 180 cm. Furthermore, those who reported using a teaching board with height of 180–190 were also at increased risk of pain in one or both elbows [41]. These heights may lead to adoption of prolonged neck extension positions when writing on or reading from the board and ultimately contribute to neck pain.

4.6 Workload

Although it is assumed that physically school teachers' work is varied and relatively light [64], research on this study population has demonstrated that they are exposed to high workloads. Rapid physical activity has been significantly associated with shoulder pain, wrists/hands pain and hips/thighs pain among school teachers in Botswana [40]. Similarly, primary school teachers in Samsun, Turkey who reported physical activity were two times at risk of neck pain when compared to those who did not report physical activity [34]. Walking up and down stairs was associated with LBP among secondary school teachers in Putrajaya, Malaysia [9].

High workload has been significantly associate with MSDs among preparatory teachers in Egypt [39]. Addis Ababa teachers who reported high work load were four times increased risk of reporting LBP than those who were not [14]. This is consistent with results of primary school teachers in Egypt where job demand was significantly associated with MSDs [50]. Physical workload has also been associated with feet pain among women teachers, nurses and sonographers in Sweden [64]. In Londrina, Brazil, high number of students in a classroom were associated with upper limbs pain [38].

The association between high job demand and MSDs might be due to the nature of teachers' work which by its nature is physically demanding. When the physical work load is reduced, the impact of job demand and onset of MSDs is reduced [65]. Apart from teaching students, teachers are also involved in lesson preparation, assessments of students' work and being involved in the extracurricular activities such as sports. Teachers also participate in different school committees. These may cause teachers to suffer adverse mental and physical health issues due to the variety of job functions [1].

4.7 Psychosocial factors

School teachers are considered to experience high level of psychological stress [64]. High psychological job demands have been associated with LBP [63], upper back pain and shoulder pain among school teachers in Botswana [40]. Similar results have been reported among secondary school teachers in Malaysia where those who reported high psychological job demands were at increased risk of developing LBP compared to those who reported low psychological job demands [43]. Psychological job demands have been associated with neck, shoulder, hands, lower back and feet pain among women teachers, nurses and sonographers in Sweden [64]. It has been suggested that the more psychological demands needed for a particular task, the greater the possibility to develop any kind of MSDs regardless of the body area [66].

The study of teachers working in governmental primary schools in Addis Ababa, Ethiopia found that those who reported a poor or fair work social environment were at increased risk of LBP than those who had good work environment [14]. Similarly teachers in another study conducted in Gondar town in Ethiopia, who reported to have stress were more likely to report LBP than those who did not report stress [57].

Mild to moderate and severe to extremely severe stress have been associated with experiencing LBP and NSP among secondary school teachers in Malaysia [43].

High anxiety and very low colleague support have been associated with MSDs among preparatory teachers in Egypt [39]. Parallels could be drawn to results of Malaysia secondary school teachers who were found to be at increased risk of LBP and NSP due to mild to moderate and severe to extremely severe anxiety [43].

Teachers in Tehran, Iran who reported low job satisfaction were more likely to develop low back when compared to those who reported high job satisfaction [56].

Low skill discretion and low supervisor support have been significantly associated with reporting low back and neck and/or shoulder pain among Malay teachers, respectively [43].

5. Protective factors

Some factors have shown a protective effect against MSDs among school teachers. These include factors such as workplace support, regular physical exercise and perceived better health.

5.1 Workplace support

A protective effect was demonstrated for Botswana teachers who reported high supervisor support. These teachers were less likely to report neck, upper back pain and hip/thigh pain as compared to those who reported low supervisor support [40]. High supervisor support was also a protective factor against LBP among Kenyan teachers [24]. Ethiopian teachers in Amhara region who reported satisfaction with work environment and culture were showed decreased odds for reporting LBP and those who had an office were also less likely to report LBP [57].

Nursery and primary school teachers who reported that there were three of them per class were less likely to report upper back, low back pain and pain on one or both ankles or feet [41].

In a study carried out among women teachers, nurses and sonographers, those who reported high job control were less likely to report shoulder, hands, lower back and feet pain. The study further demonstrated that those in leadership were less likely to report neck, shoulder and lower back pain [64].

5.2 Regular physical exercise

Physical exercise of more than 5 h per week was associated with reduced odds of reporting upper back pain [40] and LBP [63] among school teachers in Botswana. This was in comparison to teachers who reported five or less hours of weekly exercise. Similarly, teachers in Amhara region in Ethiopia who exercised were less likely to report LBP compared to those who did not exercise [57]. Chinese school teachers who reported exercising for seven or more hours per week were less likely to experience NSP compared to who exercise for less than 7 h per week [44]. Parallels could also be drawn to the results of a study conducted among Ethiopian teachers where those who exercised reported decreased odd of NSP compared to those who did not exercise [13]. Physical exercise was also associated with decreased odds of reporting neck pain among Iranian teachers [67], and upper back pain and LBP among school teachers in Enugu State, Nigeria [41]. Exercise habits also had a protective effect against neck and upper extremity pain among teachers in Turkey [51]. Saudi teachers who reported that they were involved in sports were less likely to report foot pain [60].

5.3 Perceived better health

Teachers who were generally healthy in a study conducted in Iran, were found to be less likely to experience neck pain [67]. Better self-perceived mental health reported by Malay teachers demonstrated a protective effect against LBP and NSP [43].

6. Management of MSDs

As reflected above, MSDs are common among teachers with different contributing factors. This means that management of these need to evaluate risk factors for MSDs carefully before coming up with interventions to address them. The interventions for disorders need to be contextualized for them to be effective and to take into consideration, the risk factors for these conditions and how they interact with each other. MSDs disproportionately affect females compared to their male counterparts among school teachers in Saudi Arabia [23]. Aging and improper postures have also contributed to experience of MSDs among teachers in Punjab [68].

6.1 Workplace preventative programmes

Preventive programmes are required for management of these disorders and this should ideally be at organizational level rather than individual level [7] and also use health promotion approaches to them prevent repetitive strain injury (RSI) [33]. Previous research speak to prevention measures to reduce back pain [24]. In a study conducted among Chinese teachers, there was a statistically significant improvement in attitudes, awareness, symptoms on neck and back pain after 6 and 12 months post intervention. Researchers had administered a multi-faceted workplace intervention comprising of health education through lectures, workplace ergonomic training and public awareness materials using posters and brochures and assessed pre and post intervention effects of the workplace programme [28]. Similar findings were reported among nursery school teachers who underwent an Extension oriented exercise programme to prevent LBP in nursery school teachers. The programme was found to alleviate LBP among teachers who received brochure and exercise programme done by a physiotherapist compared to those who received only thc brochure [33]. Preventive interventions focusing on posture have also been shown to work including reducing amount of time on awkward postures such as knee bending among pre-school teachers in Germany [69].

6.2 Individual coping strategies

Individual coping strategies are an important consideration for managing MSDs among teachers. This is because teachers may engage or prefer certain self-help therapies which may not necessarily be effective in prevention and management of MSDs as reported in one study where teachers used thermal spring therapy and/or painkillers to cope with their MSD pain [70]. Coping mechanism used by people suffering from MSDs can be influenced by gender and social class. This has been shown in one study where men in lower social class were found to prefer avoidant coping mechanisms compared to female counterparts while females in low social class used less problem solving methods to cope with MSDs [71]. Maintaining regular physical activity both at work and at home has demonstrated benefits for physical education teachers in terms of experiencing MSDs and their overall health and wellbeing especially their cardiovascular health [72] indicating a need to approach management of MSDs using a comprehensive risk based approach.

6.3 A comprehensive model of MSDs at work

The importance of a comprehensive multi-faceted programme to tackle MSDs is underlined by findings from a systematic review which demonstrated that massage therapy alone, a common mode of treatment for MSDs had limited benefits among patients with neck and back pain and no statistically significant benefits when compared to other treatments [73]. Additionally, patients diagnosed with MSDs commonly have other medical conditions including mental health and gastrointestinal conditions with those patients more likely to report a severe form of MSDs [74]. These patients may benefit from a holistic approach to the MSDs. A model of managing MSDs therefore requires further exploration and consideration. This is the bio-psychosocial (BPS) model of pain management.

6.3.1 The bio-psychosocial model

The model consists of three factors operating in the patient's life which are the environment, biological and cognitive factors [75]. The model posts that biomedical approaches alone have not been effective in managing pain especially MSDs and therefore a more holistic approach consisting of managing the patient biological factors, their environment and the way they think about pain, cognitive factors is key.

The model brings together an appreciation of how the risk factors already alluded to can all be incorporated effectively in a workplace programme for prevention and management of MSDs. The risk factors can seem to be too many and overwhelming to tackle for both patient and healthcare providers. However by designing a programme consisting of biological risk factors relating to the person and disease itself, cognitive factors and the environment in which illness occurs. The BPS model emphasizes the importance of making health within the patients' context taking into consideration the patients' sociocultural beliefs about illness, worries and concerns they may have about the meaning of the illness for their job and them as a person and possible coping mechanism that they have which may enhance or detract from effective interventions. Likely benefits of the bio-psychosocial model are echoed by Waddell who estimates that incapacity and sickness absence from these disorders could be reduced by up to 50% [76] and expounds on the key components of the BPS model of disability.

7. The impacts of MSDs on teachers

An important question to consider is why the public, the employers and employees must be concerned about effective prevention and management of MSDs in general and more importantly among teachers. There are bound to be costs encountered by the employee, the employer and the public because of the employees' inability to perform their duties effectively. The costs can be both tangible and intangible, direct and indirect. Assessing the impact of MSDs must therefore take a 360° view of who is affected and how are they affected. The impact can be felt at individual level and societal level.

7.1 Individual level impact

MSDs present with symptoms of pain, fatigue and functional limitations [71]. They have also been shown to adversely affect the physical and emotional components of quality of life and a likely cause of future ill health and disability [26]. Work performance is another impact of MSDs shown to affect academic teachers due to lack of adequate mental and physical rest from work even while at home [77]. A study among teachers in Botswana also noted that the effect of MSDs included functional limitations and at times career change with important implications for limited resources [40]. Increased sick leave among female teachers as well as rising levels of depression were found to be associated with having MSDs in Turkey [70]. It is clear that with symptoms experienced from these disorders, individual teachers are likely to experience functional limitations at home and at work.

7.2 Economic impact

Treatment of MSDs in general has been shown to provide economic benefits in terms of keeping people employed and earning an income in addition to reducing sickness absence from work. A study done among adults with MSDs in the UK, where 54% were employed, it showed that an average of 3.8 days were lost due to work absenteeism. The study further found that reduced functional limitations led to a reduction in the patients' ability to remain in employment, higher chances of claiming disability benefits and sickness absence [78], which all add to the economic costs of MSDs. Patients experiencing MSDs in another UK study reported that their MSDs contributed significantly to their inability to work (74%) with a quarter reporting inability to find a suitable job because of the MSDs, low job satisfaction (68%) and half experiencing limited career choices and similar proportion experiencing reduced household income [74]. These were however not all teachers but it is likely that even teachers would experience similar challenges given the physical and psychological demands of their role as teachers and the nature of MSDs. These economic cost of MSDs has long been established. Canada reported an estimated economic cost due to MSDs as early as two decades ago at 26 billion Canadian dollars with the bulk of the costs being due to healthcare resource utilization and disease sequelae [79].

7.3 Societal level impact

The impact of MSDs among teachers on other aspects of the society are not well studied. For example, the impact on the pass rates of their students, career choices of their students and psychological wellbeing of the students. It is expected that students will experience some anxiety related to sickness absence of their teacher or having a temporary teacher to replace their substantive teacher. It is also possible that the temporary teacher may not have the same qualifications and experience as the substantive teacher, they may also not have the organizational context or institutional memory which would help them to navigate the school environment and the so called difficult students effectively. This is an important area of future research.

8. Conclusion

Although self-administered questionnaires have recall biasness, introspective inability and may be subject to wrong interpretation of questions, they have been commonly used to investigate MSDs. This chapter demonstrated that MSDs are common among teachers despite of their geographical location. Additionally, MSDs are a cause of pain and suffering for teachers globally. Some countries have higher than average prevalence rates for all MSDs. Physical and psychosocial risk factors have been associated with MSDs of different body regions. Factors such as high supervisor support, high job control and regular physical exercise have been shown to have a protective effect against MSDs. Due to the effects of MSDs on individual life, work attendance and productivity it is important to manage these. Because MSDs tend to affect more than one body site and are mediated by multiple factors, a workplace approach to managing these should be holistic and as comprehensive as reasonably practicable. Future research using longitudinal study designs should be conducted to establish the casual effect of work-related and psychosocial factors in development of MSDs. Research is also needed to identify innovations that can reduce the prevalence of these disorders.

Conflict of interest

The authors declare no conflict of interest.

Author details

Patience Erick[1*], Tshephang Tumoyagae[1] and Tiny Masupe[2]

1 Department of Environmental Health, University of Botswana, Gaborone, Botswana

2 Department of Family Medicine and Public Health, University of Botswana, Gaborone, Botswana

*Address all correspondence to: erickp@ub.ac.bw

References

[1] Erick PN, Smith DR. A systematic review of musculoskeletal disorders among school teachers. BMC Musculoskeletal Disorders. 2011;**12**:260

[2] Sealetsa OJ, Thatcher A. Ergonomics issues among sewing machine operators in the textile manufacturing industry in Botswana. Work. 2011;**38**(3):279-289

[3] Coledam DHC, da Silva YM. Predictors of health-related absenteeism, presenteeism and sick leave among Brazilian elementary school teachers: A cross-sectional study. Work. 2020;**67**(3):709-719

[4] Punnett L, Wegman DH. Work-related musculoskeletal disorders: The epidemiologic evidence and the debate. Journal of Electromyography and Kinesiology. 2004;**14**(1):13-23

[5] Cardoso JP et al. Prevalence of musculoskeletal pain among teachers. Revista Brasileira de Epidemiologia. 2009;**12**(4):604-614

[6] Chong EY, Chan AH. Subjective health complaints of teachers from primary and secondary schools in Hong Kong. International Journal of Occupational Safety and Ergonomics. 2010;**16**(1):23-39

[7] da Silva NR, Almeida MA. Physical and postural aspects of teachers during work activity. Work. 2012;**41**(Suppl 1): 3657-3662

[8] Amit LM, Malabarbas GT. Prevalence and risk-factors of musculoskeletal disorders among provincial high school teachers in the Philippines. Journal of UOEH. 2020;**42**(2):151-160

[9] Nur Farahwahida MA et al. Work task and job satisfaction predicting low back pain among secondary school teachers in Putrajaya. Iranian Journal of Public Health. 2016;**45**(Supple. 1):85-92

[10] Musa-Olokuta A, Syed Q. Ergonomics study of the incidence of musculoskeletal disorder among the school teachers in Egba division of Ogun State Nigeria. Journal of Science and Technology Research. 2020;**2**(1):13-20

[11] Aldukhayel A et al. Musculoskeletal pain among school teachers in Qassim, Saudi Arabia: Prevalence, pattern, and its risk factors. Cureus. 2021;**13**(8):e17510

[12] Karimian R et al. Association between upper-extremity musculoskeletal disorders and upper cross syndrome among teachers, and the effects of NASM corrective exercises along with ergonomic intervention on their upper-extremity musculoskeletal disorders. Studies in Medical Sciences. 2021;**3**(10):753-763

[13] Temesgen MH et al. Burden of shoulder and/neck pain among school teachers in Ethiopia. BMC Musculoskeletal Disorders. 2019;**20**(1):1-9

[14] Tsega Ab A et al. Prevalence and associated factors of low back pain among teachers working at governmental primary schools in Addis Ababa, Ethiopia: A Cross Sectional Study. Biomedical Journal of Scientific & Technical Research. 2018;**10**(1):7516-7521

[15] Cohen L, Manion L, Morrison K. Research Methods in Education. 8th ed. London: Routledge; 2017

[16] Burgess TF. Guide to the Design of Questionnaires. Leeds: University of the Leeds; 2001

[17] Assunção A, Abreu MNS. Pressure to work, health status, and work conditions of schoolteachers in Basic Education in Brazil. Cadernos de Saúde Pública. 2019;**35**(Suppl. 1):e00169517

[18] Macklin Mary S. Effectiveness of McKenzie Therapy on Low Back Pain among School Teachers in Selected Schools at Kanyakumari District. Marthandam: Medical University; 2018. p. 103

[19] Santos MCS et al. Association between chronic pain and leisure time physical activity and sedentary behavior in school teachers. Behavioral Medicine. 2018;**44**(4):335-343

[20] Chiu TT, Lam PK. The prevalence of and risk factors for neck pain and upper limb pain among secondary school teachers in Hong Kong. Journal of Occupational Rehabilitation. 2007;**17**(1):19-32

[21] Assiri A, Al-Musa HM. Assessment of health status of male teachers in Abha City, Saudi Arabia. Middle East Journal of Family Medicine. 2015;**13**(7):22-29

[22] Tsuboi H et al. Psychosocial factors related to low back pain among school personnel in Nagoya, Japan. Industrial Health. 2002;**40**(3):266-271

[23] Althomali OW et al. Prevalence and factors associated with musculoskeletal disorders among secondary schoolteachers in Hail, Saudi Arabia: A Cross-Sectional Survey. International Journal of Environmental Research and Public Health. 2021;**18**(12):6632

[24] Elias HE, Downing R, Mwangi A. Low back pain among primary school teachers in Rural Kenya: Prevalence and contributing factors. African Journal of Primary Health Care Family Medicine. 2019;**11**(1):e1-e7

[25] Nyawose ZZ, Naidoo R. Prevalence of shoulder musculoskeletal disorders among school teachers: A systematic review. South African Journal for Research in Sport, Physical Education and Recreation. 2019;**41**:51-61

[26] Vega-Fernandez G et al. Musculoskeletal disorders associated with quality of life and body composition in urban and rural public school teachers. Frontiers in Public Health. 2021;**9**:607318

[27] Cieza A et al. Global estimates of the need for rehabilitation based on the Global Burden of Disease study 2019: A systematic analysis for the Global Burden of Disease Study 2019. The Lancet. 2020;**396**(10267):2006-2017

[28] Shuai J et al. Assessing the effects of an educational program for the prevention of work-related musculoskeletal disorders among school teachers. BMC Public Health. 2014;**14**:1211

[29] Ng YM, Voo P, Maakip I. Psychosocial factors, depression, and musculoskeletal disorders among teachers. BMC Public Health. 2019;**19**(1):234

[30] Arshad H et al. Prevalence, pattern of musculoskeletal pain disorders and related factors among female school teachers. Pakistan Journal of Medical and Health Sciences. 2021;**15**:1923-1926

[31] Darwish MA, Al-Zuhair SZ. Musculoskeletal pain disorders among secondary school Saudi female teachers. Pain Research and Treatment. 2013;**2013**:878570

[32] Alharbi TA, Abadi S, Awadallah NJ. Prevalence and risk factors of musculoskeletal pain among governmental male secondary school teachers. Middle East Journal of Family Medicine. 2020;**7**(10):77

[33] Chaiklieng S, Suggaravetsiri P. Risk factors for repetitive strain injuries among school teachers in Thailand. Work: A Journal of Prevention Assessment & Rehabilitation. 2012;**41**:2510-2515

[34] Durmus D, Ilhanli I. Are there work-related musculoskeletal problems among teachers in Samsun, Turkey? Journal of Back and Musculoskeletal Rehabilitation. 2012;**25**(1):5-12

[35] Vaghela NP, Parekh SK. Prevalence of the musculoskeletal disorder among school teachers. National Journal Physiology, Pharmacy and Pharmacology. 2018;**8**(2):197-201

[36] Alias AN et al. Prevalence of musculoskeletal disorders (MSDS) among primary school female teachers in Terengganu, Malaysia. International Journal of Industrial Ergonomics. 2020;**77**:102957

[37] Solis-Soto MT et al. Prevalence of musculoskeletal disorders among school teachers from urban and rural areas in Chuquisaca, Bolivia: A cross-sectional study. BMC Musculoskeletal Disorders. 2017;**18**(1):425

[38] Gabani FL et al. Chronic musculoskeletal pain and occupational aspects among Brazilian teachers. International Journal of Occupational Safety and Ergonomics. 2022;**28**(2):1304-1310

[39] Korish MM. Work related musculoskeletal disorders among preparatory school teachers in Egypt. Egyptian Journal of Occupational Medicine. 2017;**41**(1):115-126

[40] Erick P, Smith D. The prevalence and risk factors for musculoskeletal disorders among school teachers in Botswana. Occupational Medicine & Health Affairs. 2014;**2**(178):1-13

[41] Ojukwu CP et al. Prevalence, pattern and correlates of work-related musculoskeletal disorders among school teachers in Enugu, Nigeria. International Journal of Occupational Safety and Ergonomics. 2021;**27**(1):267-277

[42] Eggers LS, Pillay JD, Govender N. Musculoskeletal pain among school teachers: Are we underestimating its impact? Occupational Health Southern Africa. 2018;**24**(2):46-50

[43] Zamri EN, Moy FM, Hoe VC. Association of psychological distress and work psychosocial factors with self-reported musculoskeletal pain among secondary school teachers in Malaysia. PLoS One. 2017;**12**(2):e0172195

[44] Yue P, Liu F, Li L. Neck/shoulder pain and low back pain among school teachers in China, prevalence and risk factors. BMC Public Health. 2012;**12**:789

[45] Converso D et al. Musculoskeletal disorders among preschool teachers: Analyzing the relationships among relational demands, work meaning, and intention to leave the job. BMC Musculoskeletal Disorders. 2018;**19**(1):156

[46] Lawrence I. Musculoskeletal disorders in Nigeria nursery schools: Work-related risk reduced. Advances in Life Science and Technology. 2012;**5**:12-15

[47] Abdulmonem A et al. The prevalence of musculoskeletal pain & its associated factors among female Saudi school teachers. Pakistan Journal of Medical Sciences. 2014;**30**(6):1191-1196

[48] Kraemer K, Moreira MF, Guimarães B. Musculoskeletal pain and ergonomic risks in teachers of a federal institution. Revista

Brasileria de Medicina do Trabalho. 2021;**18**(3):343-351

[49] Mumtaz A et al. Prevalence of shoulder pain among school teachers of Lahore, Pakistan. International Annals of Medicine. 2018;**2**(3):1-4

[50] Ebied EM. Work-related musculoskeletal pain among primary school teachers: A recommended health promotion intervention for prevention and management. World Journal of Nursing Sciences. 2015;**1**(3):54-61

[51] Baskurt F, Baskurt Z, Gelecek N. Prevalence of self-reported musculoskeletal symptoms in teachers. S.D.Ü. Sağlık Bilimleri Enstitüsü Dergisi Cilt. 2011;**2**(2):58-64

[52] Barbosa REC et al. Back pain occurred due to changes in routinary activities among Brazilian schoolteachers during the COVID-19 pandemic. International Archives of Occupational and Environmental Health. 2022;**95**(2):527-538

[53] Vidal-Conti J et al. Knowledge of low back pain among primary school teachers. International Journal of Environmental Research and Public Health. 2021;**18**(21):11306

[54] Mustafa GRA. Work-related factors associated with low back pain (LBP) among secondary and primary schoolteachers in Jordan. Psychology and Education. 2020;**57**(9):1983-1987

[55] Kebede A et al. Low back pain and associated factors among primary school teachers in Mekele City, North Ethiopia: A cross-sectional study. Occupational Therapy International. 2019;**2019**:3862946

[56] Bandpei MMA et al. Occupational low back pain in primary and high school teachers: Prevalence and associated factors. Journal of Manipulative and Physiological Therapeutics. 2014;**37**(9):702-708

[57] Beyen TK, Mengestu MY, Zele YT. Low back pain and associated factors among teachers in Gondar Town, North Gondar, Amhara Region, Ethiopia. Occupational Medicine & Health Affairs. 2013;**1**(5):1000127

[58] Gupta G, Sharma A. Prevalence of low back pain among higher secondary school teachers of Kanpur, India. Journal of Orthopaedics and Physiotherapy. 2018;**1**(1):103

[59] Grant KA, Habes DJ, Tepper AL. Work activities and musculoskeletal complaints among preschool workers. Applied Ergonomics. 1995;**26**(6):405-410

[60] Alqahtani TA. The prevalence of foot pain and its associated factors among Saudi school teachers in Abha sector, Saudi Arabia. Journal of Family Medicine and Primary Care. 2020;**9**(9):4641-4647

[61] Pillastrini P et al. Effectiveness of an at-work exercise program in the prevention and management of neck and low back complaints in nursery school teachers. Industrial Health. 2009;**47**(4):349-354

[62] Chand RK et al. Prevalence of musculoskeletal disorders, associated risk factors and coping strategies among secondary school teachers in Fiji. Rawal Medical Journal. 2020;**45**(2):377-381

[63] Erick PN, Smith DR. Low back pain among school teachers in Botswana, prevalence and risk factors. BMC Musculoskeletal Disorders. 2014;**15**:359

[64] Arvidsson I et al. Cross-sectional associations between occupational factors and musculoskeletal pain

in women teachers, nurses and sonographers. BMC Musculoskeletal Disorders. 2016;**17**:35

[65] Bugajska J et al. Psychological factors at work and musculoskeletal disorders: A one year prospective study. Rheumatology International. 2013;**33**(12):2975-2983

[66] Hestbaek L et al. Comorbidity with low back pain: A cross-sectional population-based survey of 12- to 22-year-olds. Spine (Phila Pa 1976). 2004;**29**(13):1483-1491

[67] Ehsani F et al. Neck pain in Iranian school teachers: Prevalence and risk factors. Journal of Bodywork and Movement Therapies. 2018;**22**(1):64-68

[68] Saleem M, Bashir MS, Noor R. Frequency of musculoskeletal pain in female teachers. Annals of King Edward Medical University;**20**(3):245

[69] Burford EM et al. The comparative analysis of postural and biomechanical parameters of preschool teachers pre- and post-intervention within the ErgoKiTa study. Ergonomics. 2017;**60**(12):1718-1729

[70] Korkmaz NC, Cavlak U, Telci EA. Musculoskeletal pain, associated risk factors and coping strategies in school teachers. Scientific Research and Essays. 2011;**6**(3):649-657

[71] Christensen U et al. Socioeconomic position and variations in coping strategies in musculoskeletal pain: A cross-sectional study of 1,287 40- and 50-year-old men and women. Journal of Rehabilitation Medicine. 2006;**38**(5):316-321

[72] Pihl E, Matsin T, Jürimäe T. Physical activity, musculoskeletal disorders and cardiovascular risk factors in male physical education teachers. The Journal of Sports Medicine and Physical Fitness. 2002;**42**(4):466-471

[73] Bervoets DC et al. Massage therapy has short-term benefits for people with common musculoskeletal disorders compared to no treatment: A systematic review. Journal of Physiotherapy. 2015;**61**(3):106-116

[74] Zheltoukhova K, O'Dea L, Bevan S. Taking the Strain: The Impact of Musculoskeletal Disorders on Work and Home Life. Lancaster: Lancaster University; 2012. p. 60

[75] Keefe FJ et al. Pain in arthritis and musculoskeletal disorders: The role of coping skills training and exercise interventions. The Journal of Orthopaedic and Sports Physical Therapy. 1996;**24**(4):279-290

[76] Waddell G. Preventing incapacity in people with musculoskeletal disorders. British Medical Bulletin. 2006;**77-78**:55-69

[77] Niciejewska M, Kasian S. Musculoskeletal disorders related to the professional work of academic teachers and the quality of their work. Quality Production Improvement. 2019;**1**(1):47-54

[78] Dall TM et al. Modeling the indirect economic implications of musculoskeletal disorders and treatment. Cost Effectiveness and Resources Allocation. 2013;**11**(1):5

[79] Coyte PC et al. The economic cost of musculoskeletal disorders in Canada. Arthritis Care and Research. 1998;**11**(5):315-325

Chapter 4

Cold Water Exposure for Maritime Workers: A Scoping Review

Emily Walsh and Heather Carnahan

Abstract

For many of those working in maritime industries, it is very common to be exposed to harsh environments, such as cold water, on a regular basis. We conducted a scoping review on peer reviewed, published papers to summarize the literature on the topic cold water exposure and non-freezing cold water injuries in the maritime industries. First, industry experts were consulted, then a PICO model was created to define the search terms for the review. The initial search produced 690 abstract. Of these abstracts, 14 were considered to be relevant to the review. The scoping review findings illustrated the lack of research that currently exists in relation to cold water exposure in the maritime industries. Within the available, albeit limited, literature, evidence suggests that there are several cold-water injuries that occur in the maritime industries. These include occupational dermatosis, Raynaud's phenomena, finger blanching, and hand numbness. Performance decrements were also reported. The current gaps include a lack of documentation of minor and non-fatal injuries, the amount of exposure, and training protocols. There is a need to improve cold-water training regulations for those working in the maritime industries and for proper injury documentation, both of which can significantly benefit safety.

Keywords: cold-water, safety, maritime industry, cold exposure, injury

1. Introduction

Occupational health and safety is not a new concern, especially for those who work in the maritime industry. Many of the people who work in this industry deal with heavy equipment, risky situations, and various technologies on a regular basis, all the while working in harsh conditions. Cold water is prevalent in these environments, which workers are routinely exposed to. It is important to understand the relationship that the maritime industry has with cold exposure and its effects on workers to ensure that long term health and safety is not being compromised. This scoping review explored the literature on cold exposure in the maritime industries, and its impact on maritime workers.

There are two main types of cold exposure that workers in the maritime industry are exposed to: acute and chronic. Acute occurs when an individual is exposed to cold conditions or an environment once, for a limited period of time. On the other hand, chronic refers to these conditions repeatedly for an extended period of time. Chronic exposure is the primary type that those in the maritime industry experience, however, both types can lead to an individual experiencing a cold exposure injury while working [1]. These injuries can be broken down into two different categories as well. The most common injury people are familiar with is frostbite, an example of a freezing cold exposure injury that occurs below freezing conditions for an extended period of time. However, it is also possible to experience cold exposure injuries about the freezing point, which are considered to be non-freezing [1]. A non-freezing cold exposure injury occurs from prolonged exposure to wet, cold, non-freezing (between 0 and 15°C) conditions [1]. These injuries generally involve the soft tissues, nerves, and vasculature of the distal extremities, such as the hands and feet. Both cold exposure and its related injuries are concerning because little is understood about the consequences of chronic exposure to cold conditions.

Current literature regarding cold water exposure is primarily limited to research regarding acute instances, including studies conducted by the military. It is strongly supported that acute exposure to cold water can lead to significant deficits in tactile sensitivity, nerve function and motor performance [2–4]. In fact, following acute exposure to 2°C water, participants experienced impairments in both tactile sensitivity and manual dexterity in as little as 90 seconds [4]. Similar decrements in fine and gross manual dexterity also occurred following acute exposure to 10°C water [5].

As stated previously, individuals who work in the maritime industry experience chronic cold-water exposure. Due to this chronic exposure, do individuals habituate to performance in cold conditions? One study indicated that training in cold conditions may improve an individual's ability to work in the cold [6]. Results showed that although the time it took to complete a peg board test did not improve, but participants' accuracy did improve [6]. However, this study did not address the effects of chronic cold-water exposure, such as seen in an occupational setting.

A recent study examined how chronic cold-water exposure impacted tactile sensitivity and motor performance following acute exposure to cold conditions in a group of fish harvesters [7]. The results suggest that individuals with this chronic exposure may experience impaired sensory performance due to nerve impairments in their hands, indicating that rather than developing adaptations to the cold, the fish harvesters experienced non-freezing cold exposure injuries [7]. Sensory deficits related to cold exposure, such as those just reported, often occur over gradually over time rather than all at once. This allows the deficit to go undetected until its effects are more prominent. The sensory findings from this study were unexpected, which raised further questions surrounding the effects of cold water exposure in the maritime industry. Effects such as these can have a serious impact on the abilities and safety of those working in this industry, which led the authors to consider if there are other studies that have investigated this. It was this study that led to the development of this scoping review. Are these exposures being documented in the maritime industry by other research teams?

In order to explore this topic, the authors conducted a literature review. Due to the anticipated limited amount of available literature, a scoping review was determined to be the best methodology for this review paper. A scoping review is a type of review paper that is used to map the existing literature in a field, and is especially useful when the topic has not been extensively reviewed or documented [8]. A scoping review can summarize and disseminate the available literature and identify gaps in the research [9]. There are currently no existing published scoping reviews regarding the documentation of cold exposure in the maritime industry.

The primary purpose of this scoping review was to examine what has been documented regarding cold exposure in the maritime industry in relation to injuries, training, and general experiences. The results of this review will help direct the focus of future research projects that address the gaps in this area of literature. Our aim is for future research to prioritize both the understanding and improvement of the long-term health and safety of those who work in the maritime industries.

2. Methods

The objectives, inclusion criteria, and methods for this scoping review were specified in advance and documented in a review protocol [10]. Prior to conducting this scoping review, multiple experts from the maritime industry were consulted. This was to ensure that the research questions, and goal of the scoping review, were both timely and industry-relevant. These professionals included a commercial fish harvester, two aquaculture industry professionals, four ship captains, and a former oil and gas industry professional.

In line with the guidelines discussed by Peters and colleagues, two reviewers were involved with this scoping review and developed the scoping review protocol [8]. The methodology for this scoping review followed the standardized framework that was created by Arksey and O'Malley, including advancements made by Levac and colleagues [9, 11]. The following steps were taken for this review: (1) identifying the research question, (2) identifying relevant studies, (3) study selection, (4) charting the data, and (5) collating, summarizing, and reporting the results.

2.1 Research question

In order to address the primary purpose of this scoping review, several research questions were chosen. The research questions for this scoping review were as follows:

1. Are cold exposures in the maritime industry being documented? If so, how?

2. What training protocols related to cold exposure are documented?

3. What injuries have been reported as a result of cold exposure?

2.2 Inclusion criteria

Once the research question was developed, and using the feedback from industry experts, a PICO model was developed to define the search terms and inclusion criteria for the review (see **Table 1**). The inclusion criteria for the population and context of the scoping review was purposely chosen to be general in order to capture the larger

Population	All genders, global, maritime industry
Intervention	Working on cold water in the maritime industry
Comparison	Working on warm water in the maritime industry
Outcomes	Cold water, safety, injury, exposure, comfort, health

Table 1.
PICO model developed as part of the scoping review protocol to define inclusion criteria.

picture of the maritime industry. This review was limited to English-language, peer reviewed, published literature.

2.3 Data sources and search strategy

The initial search for this review was conducted on December 21, 2020. The two electronic databases used for the search were MEDLINE/PubMed (biomedical sciences, 1946-present) and Scopus (1823-present) [12]. The search terms used during the database searches were extracted from the PICO model as seen in in **Table 1**. These search terms included, but were not limited to: cold, cold exposure, hand, extremities, cold temperature, and fisher. Limits were set to ensure the papers would be English-language and that they were specifically related to humans. A librarian from the Marine Institute assisted in the database searches to ensure proper protocol was followed. Manual searches were also conducted using Google Scholar and the Memorial University Library. Following the initial search, multiple manual searches were conducted to determine if new relevant research had been published. This included a search of the reference lists of all relevant reviewed papers.

2.4 Eligibility criteria

The relevance of the papers pulled for the scoping review were evaluated at multiple screening stages. The first level of screening included evaluating the title and abstract of the paper to determine if the paper was related to the maritime industry, health and safety within the maritime industry, and cold exposure. Abstracts that met at least one of those criteria moved on to the next stage, in which the full papers were reviewed. During this screening stage, papers were removed from the review process if they did not specifically address cold exposure in the maritime industry.

2.5 Extraction of results

The data from the relevant papers was next compiled into a chart. This charting table served to analyze the information in the relevant papers and draw conclusions. The information from this chart was used to develop the results of this scoping review.

3. Results

The initial database searches and follow-up manual searches drew a total of 694 abstracts (excluding duplicates), in which 35 abstracts moved on to the next screening stage. During this stage, full papers were studied to determine if they were relevant to the review. For example, if the article discussed training and/or safety in the maritime industry but did not specifically address cold (water) exposure, it was excluded during this stage. Following these screening stages, 12 papers were determined to be eligible for the scoping review. All papers included in the review were peer-reviewed, journal articles. **Figure 1** shows the screening process of how 694 papers were screened down to 12 papers for this scoping review. While this scoping review encompassed the maritime industry as a whole, all of the papers deemed relevant to the review came from either the fish processing or fish harvesting sectors. Of the 12 papers included, 5 came from the fish processing sector and 3 came from the fish harvesting sector.

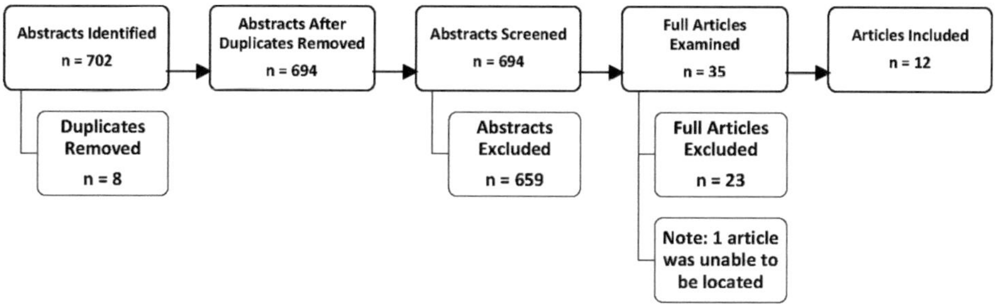

Figure 1.
Flow diagram of screening process for the scoping review.

4. Discussion

4.1 Overview of results

The key finding from this review is that there is a clear lack of cold exposure research specifically related to the maritime industry. As documented in the results, of the 12 papers that were deemed relevant for the review, most of them came from the fish processing and harvesting sectors (n = 5 and n = 3, respectively). 1 paper involved the oil and gas industry, while 3 were based on the general oil and gas industry. The fish processing sector was originally excluded from the scoping review, as seafarers (those who work on the water) were the primary targets for the review. However, due to the volume of papers that arose from the fish processing industry, it became clear that these papers were vital to the review. Unlike what the authors had expected, there was no documentation related to cold exposure training. Due to this low number of relevant papers, it is clear that there is a lack of cold exposure literature related to the maritime industry.

Those in the fish processing and harvesting sectors routinely expose their hands to cold water. For fish processors, common job tasks include cleaning of fish, packing fish for cold storage, peeling shells, cutting squid, and inspecting fish [13, 14]. While the environmental conditions for fish harvesters vary depending on the time of year, and can be unpredictable, air temperatures in fish processing plants have to be kept between 5 and 15°C since the products must be stored in cold conditions at all times [13, 14].

4.2 Cold exposure injuries

Over half the papers (n = 8) within the review documented cold exposure injuries, phenomenon, and performance deficits. While many of the injuries documented in this review are considered to be minor, several of them can have long lasting effects on an individual. For example, one paper on fish processors suggests that moderate cold exposure may be a cofactor in the development of chronic problems with muscles and joints [14]. Other studies from the fish processing industry indicate injuries such as blanching of fingers, Raynaud's phenomenon, decrease in skin barrier function, and occupational dermatosis [13, 15, 16]. As discussed in the introduction, new evidence is also emerging suggesting that those who experience chronic exposure to cold water (in this study, fish harvesters) may experience injuries that result in sensory deficits [7].

There is still much work to be done in injury documentation, especially while considering fatal and non-fatal injury documentation. The focus is often on fatal injuries, which leaves non-fatal injuries and their specific causation to be left undocumented [17]. It was also highlighted that this incomplete reporting has left injury statistics from Maritime Authorities to be unreliable [17]. Of note, this paper was not included in the review as it did not specifically address any elements related to cold exposure. In the fish harvesting industry, the reported rates of injury, fatality, and illness are also limited by the scope and accuracy of the reporting systems [18]. Another paper outside the review states that within the fish harvesting industry, there is a need to prevent severe injuries in fish harvesters while on the docks and on commercial fishing vessels through more active safety monitoring [19]. The consistency in which the maritime industry documents injuries, fatalities, and illnesses must be improved on a global scale.

So, where should the line be drawn for injury documentation in the maritime industry? It is unrealistic to expect that every scrape or minor cut can be documented in the workplace. However, injuries that impact job performance, comfort, and safety should be documented, even if they are considered to be minor. This important for accurate injury reporting, but also for occupational health and safety concerns. As mentioned above, cold exposure (injuries) may lead to future chronic injuries or issues. Therefore, by having all previous injuries documented, researchers can further the understanding of the effects of cold exposure on injuries.

4.3 Effects of cold exposure on performance

While considering the effects of chronic cold exposure on those in the maritime industries, it is important to not only view it from an injury lens, but from a performance lens as well. It is well known that cold exposure can affect performance, including for those who work in the maritime industry. Factors affecting performance (such as fatigue, discomfort, and stress) can affect anyone in many different types of situations. However, performance decrements are especially dangerous for those who work in the maritime industry for multiple reasons. Seafarers often work in dangerous environments, in which a small misstep can have costly mistakes. Additionally, the maritime industry involves a large amount of hands-on work with heavy and advanced equipment. To prevent fatal, and even minor injuries, seafarers have to be consistently aware of their surroundings.

Extreme temperatures can have a severe effect on seafarers' performance [20]. Low temperatures while working may cause fatigue, decrease mental abilities and perception, increase risk of perpetual error, and decrease an individual's ability to identify external elements [20]. In this survey, extreme temperatures had the third highest factor in affecting seafarer's performance of maintenance duties, behind only workload or stress and ship motion [20]. Another study also stated that focus decreased and human error probability increased when work temperatures shifted from normal to extreme [21]. Other deficits related to working in cold environments include a loss of balance, mobility, and strength [22].

Possible cold exposure adaptations were also documented in three of the review papers. For example, the onset of cold vasodilation response was quicker in these individuals compared to their control counterparts during hand immersion in cold water in one study [23]. During another study involving fish harvesters, fish harvesters maintained higher finger temperature and heat flow from their hands compared to their control counterparts during hand immersion in cold water, and

75% of fish harvesters experienced cold-induced vasodilation [24]. Fish harvesters in both studies appeared to experience less discomfort and pain and verbally complained less of pain [23, 24]. In addition, it has been discussed that this adaptation is not just an adaptation to pain in general, but rather to pain caused by cold exposure [25].

4.4 Cold exposure training

Prior to conducting this review, an area of interest for the authors was regarding training protocols have been documented in the literature regarding cold exposure training. However, little to no information on such training has been documented. It was reported that there is a lack of longitudinal studies on occupational health and safety issues in the fish processing sector in Asia [13]. This can be broadened to say that the same is true for the global marine industry as a whole.

During the searching phase of the review, the authors found several papers that discuss the need and importance for more safety training in the maritime industries. While these papers fall outside the scope of the review, they were of interest and therefore, mentioned here. Training courses are a key way in preventing injuries and creating a safer workplace. For example, 'Safety Training & Oceanic Fishing' by suggests that training courses have resulted in individuals using more caution in respect to less severe incidents [26].

5. Implications for research and practice

This scoping review highlights the multiple gaps in maritime industry cold exposure literature. The clear lack of industry relevant research demonstrates a lack of understanding of cold water exposure in the maritime industry. There are several recommendations to make for future literature in this area. First, more literature is needed documenting workers' experiences with cold water exposure. This may include the prevalence of cold exposure in their respective occupations, their understanding of this exposure, whether it's chronic or acute exposure, and what job tasks expose them to cold water. Secondly, the development of a standardized framework in documenting cold exposure injuries in the literature will allow researchers globally to focus their work on the prevalent injuries and further their understanding of the common cold exposure injury risks in the maritime industry. The research conducted related to maritime industry cold exposure cannot be primarily limited to the fish processing and harvesting sectors. This field of research is emerging and it is critical that these gaps are addressed. The goal of our paper is to provide a clear baseline for future research projects on cold exposure in the maritime industry. Using the information in this review paper, future research can target the current gaps in the literature while being industry relevant. Minimizing these gaps will enable both the academic and industry communities to further understand how cold exposure affects maritime workers, creating opportunities to improve their long-term health and safety.

6. Conclusions

With new evidence emerging related to the effects of acute and chronic cold-water exposure, the authors conducted a scoping review to determine how it is understood and documented in the maritime industry. It is well known that many of the people who work within this industry are commonly exposed to cold (and often wet) conditions. A standardized framework was followed throughout this review, and a PICO model was developed to determine the search criteria. Additionally, the inclusion criteria for this review included English-literature, peer-reviewed, published literature. As expected, there is very little documented regarding this exposure in the literature. Only 15 documents were deemed eligible for the review, indicating a clear lack of research. While the results are limited, there are many research opportunities in this area, and this review documents several recommendations moving forward. For example, calls for standardized injury reporting and more rigid documentation have been made in various maritime sectors, including the fish harvesting industry. It is clear that furthering the understanding of cold-water exposure, creating more in-depth cold exposure training protocols, and improving cold water exposure knowledge is needed. Reducing all types of safety risks, especially those related to cold-water, and increasing transparency and knowledge on the subject will serve to improve the long-term safety in the maritime industry.

Acknowledgements

We would like to thank and acknowledge the librarian at the Fisheries & Marine Institute of Memorial University, Catherine Lawton, for her assistance in conducting the database searches for this review.

Conflicts of interest

The authors declare no conflict of interest.

Author details

Emily Walsh and Heather Carnahan*
Fisheries and Marine Institute of Memorial University, St. John's, Canada

*Address all correspondence to: heather.carnahan@mi.mun.ca

References

[1] Heil K, Thomas R, Robertson G, Porter A, Milner M, Wood A. Freezing and non-freezing cold weather injuries: A systematic review. British Medical Bulletin. 2016;**117**:79-93

[2] Heus R, Daanen HA, Havenith G. Physiological criteria for functioning of hands in the cold: A review. Applied Ergonomics. 1995;**26**:5-13

[3] Ray M, Sanli E, Brown R, Ennis KA, Carnahan H. The combined effect of cold and moisture on manual performance. Human Factors. 2017;**60**:92-100

[4] Ray M, Power C, Luscombe T, Jones A, Carnahan H. A timeline for hand function following exposure to 2 °C water. International Journal of Industrial Ergonomics. 2019;**72**:112-118

[5] Cheung S, Montie D, White M, Behm D. Changes in manual dexterity following short- term hand and forearm immersion in 10° C. Aerospace Medical Association. 2003;**77**(4):990-993

[6] King M, Ray M, Mulligan D, Carnahan H. Does training in the cold improve cold performance? International Journal of Industrial Ergonomics. 2020;**76**:102926

[7] Armstrong C, Holden B, Walsh B, Walsh E, Carnahan H. Effects of chronic cold water exposure in fish harvesters on sensory and motor performance and cold tolerance. International Journal of Human Factors and Ergonomics. 2020;**7**(4):314-324

[8] Peters M, Godfrey CM, Khalil H, Mcinerney P, Parker D, Soares CB. Guidance for conducting systematic scoping reviews. International Journal of Evidence-Based Healthcare. 2015;**13**(3):141-146. DOI: 10.1097/XEB.0000000000000050

[9] Arksey H, O'Malley L. Scoping studies: Towards a methodological framework. International Journal of Social Research Methodology. 2005;**8**(1):19-32. DOI: 10.1080/1364557032000119616

[10] Peters M, Godfrey CM, Khalil H, Mcinerney P, Soares CB, Parker D. Guidance for the conduct of JBI scoping reviews. In: Joanna Briggs institute Reviewer's manual. The Joanna Briggs Institute. 2017;**2017**:6-28

[11] Levac D, Colquhoun H, O'Brien KK. Scoping studies: Advancing the methodology. Implementation Science. 2010;**5**:69. DOI: 10.1186/1748-5908-5-69

[12] Pham MT, Rajic A, Greig JD, Sargeant JM, Papadopoulos A, McEwen SA. A scoping review of scoping reviews: Advancing the approach and enhancing the consistency. Wiley Online Library. 2014;**5**:371-385. DOI: 10.1002/jrsm.1123

[13] Nag PK, Nag A. Hazards and health complaints associated with fish processing activities in India- evaluation of a low-cost intervention. International Journal of Industrial Ergonomics. 2007;**37**(2):125-132. DOI: 10.1016/j.ergon.2006.10.012

[14] Lundqvist GR, Jensen PL, Solberg HE, Davidsen E. Moderate cold exposure in the Faroe fishing industry. Scandinavian Journal of Work, Environment, & Health. 1990;**16**(4):278-283. Available from: http://www.jstor.org/stable/40965804

[15] Halkier-Sorensen L, Menon GK, Elias PM, Thestrup-Pedersen K, Feingold KR. Cutaneous barrier function after cold exposure in hairless mice: A model to demonstrate how cold intereferes with barrier homeostasis among workers in the fish-processing industry. The British Journal of Dermatology. 1995;**132**(3):391-401. DOI: 10.1111/j.1365-2133.1995.tb08672.x

[16] Halkier-Sorensen L, Thestrup-Pedersen K. Skin physiological changes in employees in the fish processing industry immediately following work. A field study. Contact Dermatitis. 1991;**25**(1):19-24. DOI: 10.1111/j.1600-0536.1991.tb01767.x

[17] Jensen OC. Work related injuries in Danish fishermen. Occupational Medicine. 1996;**46**(6):414-420. DOI: 10.1093/occmed/46.6.414

[18] Windle MJS, Neis B, Bornstein S, Binkley M, Navarro P. Fishing occupational health and safety: A comparison of regulatory regimes and safety outcomes in six countries. Marine Policy. 2008;**32**(4):701-710. DOI: 10.1016/j.marpol.2007.12.003

[19] Thomas TK, Lincoln JM, Bradley H, B.J., & Conway, G.A. Is it safe on deck? Fatal and non-fatal workplace injuries among Alaskan commercial fishermen. American Journal of Industrial Medicine. 2001;**40**(6):693-702. DOI: 10.1002/ajim.10010

[20] Islam R, Faisal K, Abbassi R, Garaniya V. Human error assessment during maintenance operations of marine systems- what are the effective environmental factors? Safety Science. 2018;**107**:85-98. DOI: 10.1016/j.ssci.2018.04.011

[21] Islam R, Faisal K, Abbassi R, Garaniya V. Human error probability assessment during maintenance activities of marine systems. Safety and Health at Work. 2018;**9**(1):42-52. DOI: 10.1016/j.shaw.2017.06.008

[22] Noroozi A, Abbassi R, MacKinnon S, Khan F, Khakzad N. Effects in cold environments on human reliability assessment in offshore oil and gas facilities. Human Factors: The Journal of the Human Factors and Ergonomics Society. 2013;**56**(5):825-839. DOI: 10.1177%2F0018720813512328

[23] Krog J, Folkow B, Fox RH, Lange Andersen K. Hand circulation in the cold of Lapps and north Norwegian fishermen. Journal of Applied Physiology. 1960;**15**(4):654-658. DOI: 10.1152/jappl.1960.15.4.654

[24] LeBlanc J, Hildes JA, Héroux O. Tolerance of Gaspé fishermen to cold water. Journal of Applied Physiology. 1960;**15**(6):1031-1034. DOI: 10.1152/jappl.1960.15.6.1031

[25] LeBlanc J. Local adaptation to cold of Gaspé fishermen. Journal of Applied Physiology. 1962;**17**(6):950-952. DOI: 10.1152/jappl.1962.17.6.950

[26] Poggie JJ, Pollnac RB. Safety training and oceanic fishing. Marine Fisheries Review. 1997;**59**(2):25-28

Section 3

Organizational Ergonomics

Chapter 5

Work Design in Apparel Sector

Özlem Kaya

Abstract

The human-centered design approach can be defined as the sum of the methods and procedures that make it possible to carry out each evaluation and design intervention, starting with an awareness. It is a philosophy of intervention that places not only people's needs and expectations, but above all people at the center of the design and production process for products, environments, and systems. Human-centered design is an interactive system development approach that aims to make systems usable and useful by focusing on users, their needs and requirements, and by applying human factors/ergonomics and usability knowledge and techniques. This approach increases efficiency and productivity, improves human well-being, user satisfaction, accessibility and sustainability, and eliminates the potential negative effects of use on human health, safety, and performance. Proper design and improvements in working conditions and work organization can result in increased productivity and competitiveness. In this context, work design is an extremely important issue in the apparel industry, which is one of the largest industries in the world.

Keywords: apparel sector, work design, workstation, working condition, ergonomic workplace design

1. Introduction

Despite the transition to automation along with technological developments, enterprises still need physical manpower. In addition to the increasing mechanization due to technological developments, it is necessary to pay attention to the working conditions and satisfaction of the employee due to the necessity of the human factor in the working environment. Especially in the apparel sector, which is a labor-intensive sector despite rapid technological developments, the effects of the employees on production are great, and the layout of the working environment directly affects the productivity of the employee. The productivity of the employees in the apparel sector, which is a labor-intensive sector, greatly affects the productivity, profitability, and product quality of the enterprise.

In this context, there are various principles that can be used to design real working environments in accordance with ergonomic conditions and to enable employees to work with healthier working postures. The additional cost of an ergonomic design to be carried out in the light of these principles will be very low and insignificant when compared with the cost that will arise due to working in working environments with inadequate health and safety conditions that were not designed in the light of these main principles. Many of the problems encountered in work environments can be avoided by following current health and safety regulations and guidelines on good practice. In this context, it will be possible to evaluate the tasks in the workplace, to implement some preventive rules, and to make workplaces safer environments in this context with appropriate workstation designs.

There are risk factors for employees in a working environment that does not comply with ergonomic criteria. For this reason, the working environments in the apparel sector and all the work of the employees should be arranged in a way that will meet the ergonomic conditions and make the people the least difficult.

Problems caused by the workstation arranged without taking into account the characteristics of the human body and anthropometric values can lead to short-term or chronic health problems. In many studies, it has been shown that occupational accidents and occupational diseases are caused by incompatibility of the working machine, and this issue is not given enough attention.

In this regard, special attention should be paid to and considered the common postures found in the enterprises environment, the postures that should be considered when designing workplace products or spaces. These postures:

- Standing
- Sitting
- Reaching
- Moving [1]

Improper working postures constitute one of the primary risk factors for work-related musculoskeletal disorders, from minor back pain to severe disability. It is important to take proactive steps to assess and mitigate the problem. Therefore, early identification of work-related musculoskeletal disorders and the risk factors that cause these disorders is important. More appropriate working postures have positive effects on the musculoskeletal system, allow more effective control of working performance, and reduce occupational accidents [2].

In this context, when **Figure 1** is examined (photos taken in apparel enterprises in Turkey in 2022), it is possible to see the workplace design and working postures of machine operators and quality control employees in the apparel sector. Employees in different working postures have to work in these positions for long hours. Many situations such as excessive bending, working in the same position all the time, sitting or standing all the time affect the employees negatively. At the same time, when the photograph is examined, it is possible to say that the workplace environment is not designed ergonomically.

The workspace and equipment design features that enable employees to take a more upright posture with less trunk or neck flexion are not possible with the tools and equipment given in **Figure 1** and the chairs and tables used. In this respect, it is extremely important for enterprises to pay attention to the use of ergonomically designed tools and equipment and to the points specified in terms of employee health and operating efficiency.

Work-related factors that are related to work-related musculoskeletal disorders and accelerate the discomfort process are considered as important ergonomic risk factors. Ergonomic risk factors include heavy lifting, repetitive movements, reaching, pulling, turning, etc., long-term work and intense focusing, and inappropriate

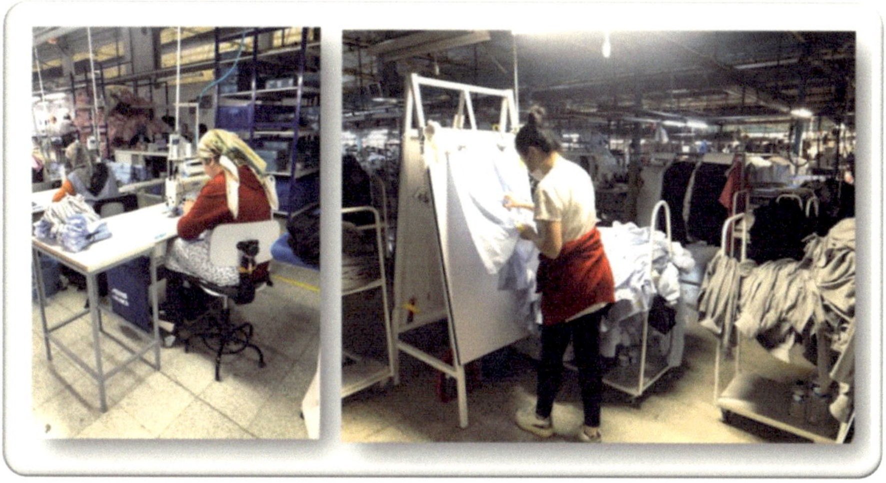

Figure 1.
Working Postures (Photos by Kaya).

Figure 2.
Working with personal protective equipment (Photos by Kaya).

working postures [3]. At the same time, personal protective equipment is another important factor in terms of risk formation. Personal protective equipment including clothing, gloves, or equipment can help minimize risk factors in the workplace. Equipment such as masks, earplugs, safety glasses, chemical aprons, safety shoes, and hard hats are also among the personal protective equipment.

When **Figure 2** is examined, it is seen that there are finger guards that must be worn while using cutting robots, and that there are employees who work with earplugs, which machine operators who work with loud noise have to use. The use of these personal protective equipment by employees is extremely important for occupational health and safety. Especially finger protectors are extremely important equipment in the prevention of work accidents that may occur when using cutting machines. In addition, the role of earplugs is extremely important in the sector where partial or long-term hearing loss is experienced as a result of long exposure to loud noise.

It is an important step for the employees to have the necessary knowledge about these equipment and to receive training on why they should be used, at the point of preventing the occurrence of risk. The correct use of personal protective equipment is an important factor that prevents the occurrence of work accidents and occupational diseases. Otherwise, there will be more risk [4, 5].

The use of personal protective equipment is important in minimizing ergonomic risks, and therefore, it should be checked whether the employees use their personal protective equipment correctly. These checks are generally intended to prevent some personal injury. It is also necessary that the controls are strictly monitored to ensure that they reduce or eliminate ergonomic risk factors and prevent the formation of new risk factors [6, 7].

When the work environment is suitable for the anatomical, physiological, psychological characteristics and capacity of the individual doing the work, harmony is achieved between the work and the person doing the work, and thus, the highest efficiency can be achieved with the least fatigue.

At the same time, in the source prepared by the ILO, edited by Hiba (1998) [8], the point especially emphasized is that improvements in working conditions and work organization can result in increased productivity and competitiveness. In this context, the work environment designs of the apparel sector should be created by considering multifaceted.

Especially the design and manufacture of clothing are an area in which very good improvements can be made. Many small industries are known to manufacture standard garments for retailers under very stringent conditions and procedures. For this reason, the improvement of these processes is very important for the sector.

When **Figure 3** is examined, it is seen that the work stations are scattered and at a level that may endanger work safety. This situation may distract the employee, as well as carry risks that may cause various injuries due to crashes and scattering.

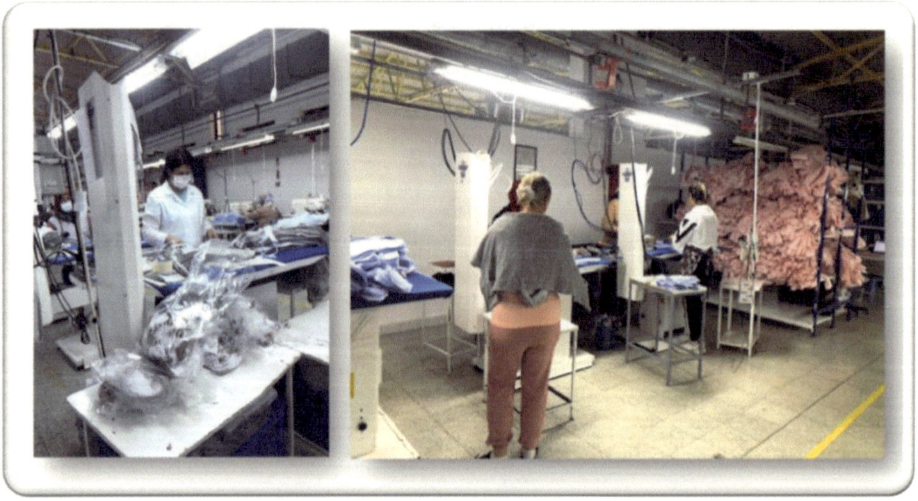

Figure 3.
Working Environments of the Apparel Sector (Photos by Kaya).

Work design plays an important role in optimizing employee performance. It is important to focus on work design to increase job satisfaction and motivation of employees. Effective work design measures the extent to which the employee is involved in tasks and tasks [9].

For these reasons, ergonomic regulation of the working environment is extremely important in reducing both employee job satisfaction and motivation and especially the problems arising from ergonomic risk factors.

Despite the developing technology, the apparel sector is a labor-intensive industry. With the increasing competitive environment, situations such as unsuitable working postures, continuous and repetitive works, inappropriate work designs, and time pressure in the apparel sector cause many problems (especially musculoskeletal system problems). In this context, ergonomics is an important tool in the prevention of work-related physical or mental disorders.

In this context, with this study, the work design processes and current situations of enterprises operating in the apparel sector have been revealed, and some sector-specific suggestions have been developed at the point of what should be done by making some determinations.

2. Work design

"Good work" is healthy and safe work in which hazards and risks are eliminated or minimized, as far as reasonably practicable. Good work is also where work design optimizes human performance, job satisfaction, and productivity.

Good work includes positive work elements that can:

- Protect employees from harm to their health, safety, and welfare.
- Improve employee health and well-being, and.
- Increase work success through higher worker productivity [10].

Work design, on the other hand, means "the content and organization of one's work tasks, activities, relationships and responsibilities" [11].

Work design also means the content, structure, and organization of tasks and activities. It is often examined in terms of work characteristics such as autonomy, workload, role issues, and feedback. Throughout history, work design has moved from focusing only on efficiency and productivity to more motivating work designs, including the social approach to work, Herzberg's two-factor model, Hackman and Oldham's work features model, Karasek's work demand control model, and the social approach to work. Like Warr's vitamin model and work, Bakker and Demerouti's resource demand model. The models make it clear that various work characteristics make up the quality of work design that benefits both employees and employers. Work design:

- Employee health and well-being,
- Attitudes such as job satisfaction and commitment,
- Employee cognitions and learning, and
- It is about productivity, absenteeism, proactivity, and innovation.

Personal characteristics of the employee play an important role in work design. They influence how employees perceive and seek specific work characteristics, help in understanding how work design exerts its impact, and have the potential to change the impact of work design.

In the literature, specific approaches to work design have tended to focus primarily on improving the design of work or enterprises processes in order to increase employee motivation and performance, although health and well-being are often considered the results of these approaches.

From the point of view of the Elements of Good Work Design, when making decisions about work tasks, activities, and responsibilities, there are usually four elements that interact together:

- Physical Elements: Aspects of the work environment or context that place physical or physiological demands on the human body.

- Biomechanical Aspects: Aspects of work including hazardous manual tasks and biomechanical risk factors leading to musculoskeletal disorders.

- Cognitive Aspects: Aspects of the work that place demands on human mental capacity.

- Psychosocial Aspects: Social, psychological, and organizational aspects of work, human capacities.

It has been consistently shown that these elements of work, summarized in **Figure 4**, have a significant impact on employees in terms of mental health, safety, well-being, and performance [12].

Billions of people spend most of their waking lives at work, so there's a good chance that work can be a positive feature of life.

Whether the work is beneficial or harmful depends largely on how it was designed. Therefore, work design is defined as the content, structure, and organization of one's tasks and activities [11, 13].

Work design researchers have considered each of these aspects when designing works, whether biological (e.g., related to noise and lift), ergonomic (e.g., lighting, information input), motivational (e.g., autonomy, diversity) or mechanical (e.g., specialization, simplification) work characteristics.

The most effective design process begins at the earliest opportunity, in the conceptual and planning stages. At this early stage, it is great luck to find ways to design hazards, to incorporate effective risk control measures and efficiencies in design **Figure 5**.

The effective design of good work takes into account: Work:

- How work is done, including the physical, mental, and emotional demands of tasks and activities;

- Task duration, frequency, and complexity; and.

- The context and systems of the work [10].

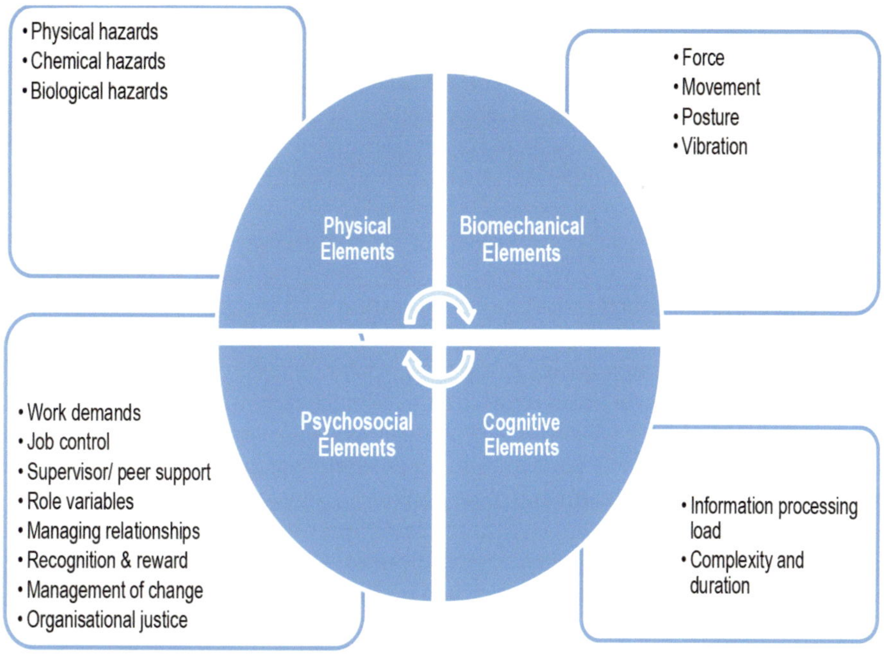

Figure 4.
Key Elements of Work Including Physical, Biomechanical, Psychosocial, and Cognitive [12].

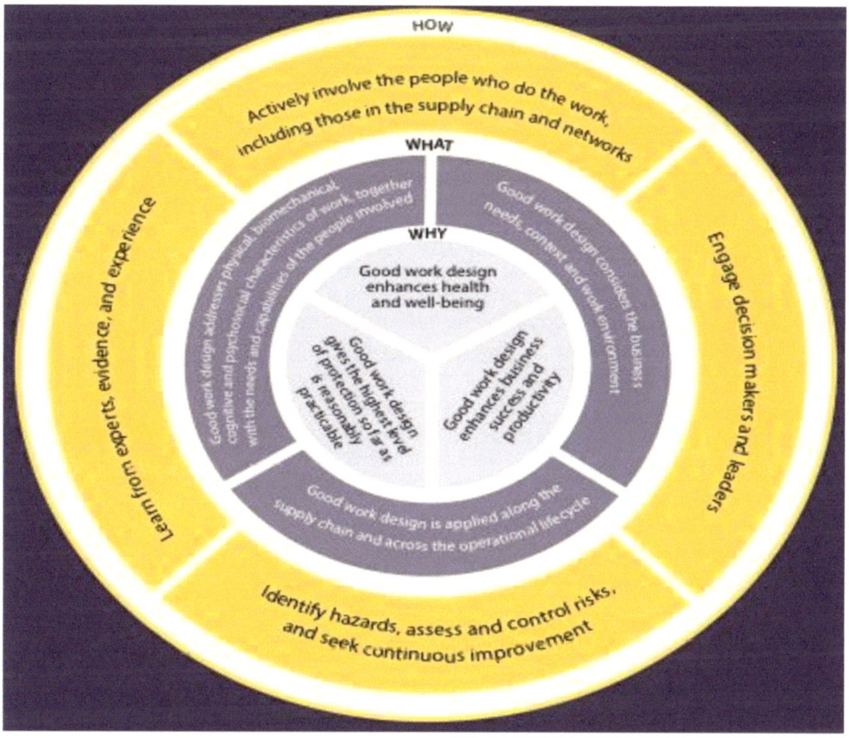

Figure 5.
Good work design principles [10].

Different outcomes of work design have been highlighted in the work design literature. While Taylor mostly focused on performance, most motivational and health-oriented work design models focused on employee well-being and attitudes. But none of the existing work design models has fully done justice to the rich empirically examined results. The immediate, i.e., individual-level outcomes of work design are grouped here in terms of health and well-being, cognitions and learning, attitudes and behaviors [14, 15].

As with the study, creation, and modification of the composition, content, structure, and environment in which works and roles are enacted, the discipline of work design plays a central role in understanding what makes work important to individuals. An integrative work design model that takes into account various work and employee characteristics in the task, social, and contextual domains, integrating research on work and team design, is extremely important at this point (**Figure 6**) [16].

Alongside this integrative work design, while all new technologies and materials present new occupational health and safety hazards, digitalization and automation have the potential to improve working conditions and the safety and health of employees in industries that have struggled for many years to protect and prevent employees from work-related risks (such as accidents, injuries, and diseases).

This potential includes:

- Laser cutters and sewing robots can take on repetitive and hazardous tasks, meaning fewer employees are exposed to hazardous chemicals or injuries associated with repetitive movements and long working hours.

- New environmentally friendly methods have been invented to eliminate the exposure of employees to silicosis instead of traditional sandblasting of jeans. Their use is also extremely important for health.

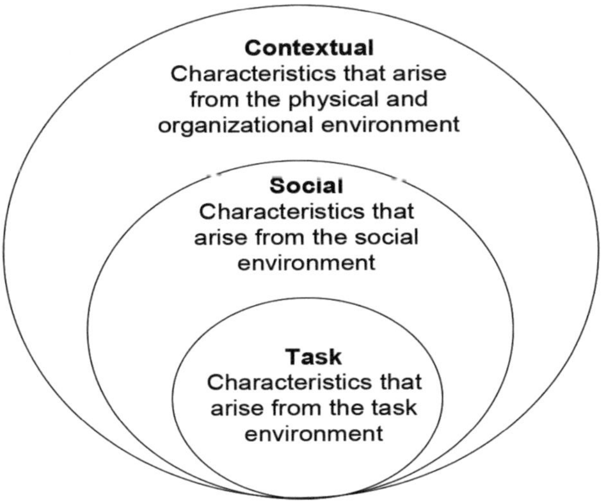

Figure 6.
Integrative framework of work design [16].

- Use of solar panels and other forms of renewable energy (can reduce reliance on highly polluting diesel-powered generators and industries can significantly reduce their use of them.)
- Resource-efficient equipment can significantly reduce industries' energy and water use and improve their environmental footprint.
- Low-cost technologies such as fire alarms, sprinkler systems, fire doors, or air conditioning can save lives and increase productivity [17].

In the light of this information, it is extremely important to make workplace designs with a more ergonomic approach for work design in the apparel sector in general and to implement this with all stakeholders, employers, and employees, and taking their opinions. In this context, the most important elements to be considered in ergonomic workplace design should be:

- Postures; especially awkward-inappropriate postures that can cause health problems. (OWAS observation method can be used for posture analysis) There are many ergonomic methods for assessing postures [18], and they differ in the body region they evaluate [19, 20].
- Working environment with working conditions such as noise, heat, humidity, lighting, air velocity [21, 22].
- The design of the production process, taking into account all the tasks performed by the employee, especially repetitive tasks. (The weight of the manufacturing parts is also an important factor.)

3. Ergonomic work design in the apparel sector

The Fourth Industrial Revolution will devastatingly change the world of labor, especially with automation and artificial intelligence. It will greatly affect employees, especially in labor-intensive manufacturing industries such as textiles and apparel, footwear, and electronics. According to the forecast of the German Smart Factories, the automation of the industrial production process will be a step toward the creation of a social network of machines and factories communicating with each other through artificial intelligence, and this will be accomplished within the next decade. This means that many jobs will be lost, with many new jobs created.

Textile and apparel manufacturing is now considered a high-risk sector of the Fourth Industrial Revolution, and the apparel sector is a major concern for government, enterprises, unions, and employees due to its high labor intensity [23].

Enterprises where apparel is produced are among the labor-intensive industries that require less capital compared with other industries in terms of their production structures. Despite the developing technologies, the apparel sector still maintains its labor-intensive structure.

In terms of the work done in the apparel sector, the employees do the works such as design, cutting, sewing, production, ironing, and packaging [24]. Due to the nature of these works, it is inevitable that there will be some work-related problems. Health problems, occupational diseases, and occupational accidents are among the frequently encountered situations, especially in terms of workstations that are not ergonomically designed and the tools used. Although the severity of these problems depends on the nature of the work, limited and continuous work postures, and very repetitive actions, it is a fact that the role of ergonomic risk factors is great. A lot of research has been done on this situation in the apparel sector, especially on working conditions and related problems [25–29]. The findings obtained in these studies revealed the effects of restricted postures, inappropriate body postures, working in the same position, repetitive movements, and especially ergonomically not designed workstations and tools used on employees.

At the same time, as Ahasan and Rabiul [30] stated in their research, the poor design of the workplace and the equipment used negatively contributes to the physical discomfort experienced by the employees. This situation may be related to the noncompliance with the labor laws and the decisions and practices of the International Labor Organization in the apparel sector in many other developed countries, especially in underdeveloped countries [31].

Recognizing and identifying ergonomic risk factors in apparel enterprises are an important first step in correcting hazards and protecting workers and improving workstations [32]. At this point, in relation to the reasons we have mentioned, work design in the apparel sector maintains its warmth in every period as an extremely important issue.

Although the apparel sector is in the less dangerous class, there are many dangers that can cause work accidents and occupational diseases in the sector. Ergonomic risks are at the forefront of these dangers, and the discomforts caused by ergonomic risks are among the priority problems in the sector. Back, waist, shoulder pain, pain in the hands, arms, elbows, burning, neck straightening, pain in the feet and legs, eye problems are the most common complaints of the employees in the sector [33]. When the apparel production stages are observed, it is seen that the employees are either constantly sitting or constantly working on their feet (**Figures 1–3**). Sitting employees work constantly by using their hands, arms, and eyes in the same position during working hours, adjustable and non-ergonomic work tables and chairs, insufficient lighting in the environment, unsuitable thermal comfort conditions, the continuity of the work, and the necessity of getting up according to the production schedule, these can cause health problems for the employees. Similar problems exist for standing employees. Especially during the whole working period, many musculoskeletal system problems occur due to working all day long directly on the hard ground. In addition, employees working in manual handling works such as carrying, stacking, and loading fabric balls on the bench are also exposed to ergonomic risks. Ergonomics is an important tool in solving the problems encountered in these workstations due to exposure or wrong doing (working postures, repetitive and continuous work, etc.) and the tools used.

Ergonomics also plays an important role in areas where conflicts between man and machine arise. It adapts the work to the person by touching different components to a single system so that each component can work in sync with the others. These components include the employee, the work environment, both physical and organizational tasks, and workspace.

In **Figure 7**, a general and specific press machine workstation image of the workstations belonging to the apparel enterprises is given.

Figure 7.
Apparel Sector Work Environment and Press Machine Workstation (Photos by Kaya).

Figure 8.
Cutting department (Photos by Kaya).

When **Figure 7** is examined, it is seen that the workstations belonging to different units in the enterprises are not ergonomically designed. When the photographs are examined, it is seen that ergonomic problems cause some problems among the employees, especially in the sewing, cutting, and ironing sections, and as a result of the observations, the incompatibility between the human and machine interface is more evident in the sewing and cutting sections. In this respect, the study focused more on these two parts.

In the cutting department (**Figure 8**), during the loading, spreading, cutting, and stacking of the fabric on the laying machine, some problems are experienced due to workstations that are not ergonomically designed, and that the height, width, cutting area of the cutting tables are not adjustable according to the employee, and some problems arise due to the inconvenience of view and operation area. Appears to be happening. At the same time, it is seen that the cutting robots, hand knives, and saw-mills used during cutting are positioned on tables that cannot be adjusted in height, and they carry risks in terms of occupational accidents due to being out of sight of the employee or very close to them.

Considering the sewing department of different work given in the photographs, it is noteworthy that there are a number of unsuitable problems in terms of sewing tables, chairs and tables used, floor surface, working area, ambient lighting, and ventilation. It is possible to solve these problems with some simple but mandatory regulations. Considering that these departments (especially the sewing department) are the departments that employ the most employees, it will be understood once again how necessary and important the work design is **Figure 9**.

Many sectoral research studies show that the sewing department draws attention as the department where the most problems are experienced in the apparel sector and where the most work development, ergonomic arrangement, and different work designs are applied. In this respect, every study that contributes to the sector is extremely important.

Figure 10 shows the non-adjustable fabric control units and the warehouse section with a dangerous design. In particular, problems such as the random laying of fabric balls on the ground and the absence of lifting tools are extremely problematic for the employee. In the other photo (**Figure 10** on the right), the warehouse section, which is not ergonomically designed, is seen. The scattered and uncontrolled parcels around the wheeled ladder create a work environment that can move at any time without a stopper and cause serious injury to the employee. At the same time, stacking products

Figure 9.
Sewing department (Photos by Kaya).

Figure 10.
Fabric Control and Warehouse Department (Photos by Kaya).

at points that are far above the access point of the employee, even with a ladder, will bring many work accidents. Therefore, as a whole, these problems unfortunately stand out as problems encountered in almost every apparel enterprises. Identifying these problems and developing an appropriate work design are extremely important for employee health and safety.

Apparel sector features that can be developed to prevent injuries include the following.

- Communication,
- Participation of employees in decision-making processes,
- Education and training of employees and management on preventive strategies, and.
- Ergonomic conditions in the factory.

To summarize briefly, the apparel sector is generally seen as a safe place to work and there are relatively few serious accidents in industries, apparel factories, compared with other industries. The hazards are different when compared with other sectors. Major health risks in this sector do not usually arise from sudden, potentially deadly hazards. Instead, the risks apparel workers face stem from more subtle hazards whose effects build up over time. Sewing machine operators face a significantly higher risk of muscle soreness and injury than employees in other occupations. Studies also show that the frequency of permanent neck and shoulder injuries increases with years of employment. As Saravanan points out (2011) [34], sewing machine operators experience as many repetitive strain injuries as data entry operators and secretaries combined. These injuries can have long-term health effects. In the light of all this information, it is once again revealed that the apparel sector, which is one of the most important sectors in the world, has the highest employment and plays an important role in the economic development of many countries, should have suitable working conditions. At the point of what needs to be done for this, it is recommended to develop work design models with the participation of the necessary experts and all stakeholders and to make working conditions more humane.

4. Conclusion

This study includes the evaluation of whether the workstations in the enterprises, the tools, and equipment used in the enterprises are ergonomic, and especially the workplace designs, which affect the health and work efficiency of the employees in the apparel sector. At the same time, the importance of work design in the sector was emphasized with the study and the fact that ergonomics awareness, which is of vital importance in work design, is very low. In this context, the importance of ergonomic information and ergonomic analysis in work design has been emphasized once again with the study, and its effect on employees has been evaluated specifically for the enterprises where the application is made and the photos are taken.

It is seen that there is a need for a wide area of development in work design, machine layout, use of equipment, and working conditions in order to provide maximum comfort to the employees in order to increase the health and well-being of the employees. For this reason, the following suggestions are presented to enterprises in the apparel sector:

In areas where work is intense such as working in inappropriate positions (cutting, production, quality control, packaging), repetitive movements (cutting, production, quality control, packaging), carrying by hand (storage, transportation), sitting, standing constantly, whole body and hand-arm vibration, especially:

- Employees should be provided with a chair or stool where they can sit at certain intervals. Care should be taken to ensure that these chairs and stools are also ergonomic.

- The work table should be adjustable for work at different heights. If it is not possible to adjust the work area, the desk should be raised using the support for tall employees. Short employees should be provided to work on a platform.

- Foot rest supports should be used. These supports will prevent the feeling of pain in the feet and allow the employee to change position. Thus, complaints of pain in the back and legs will be reduced.

- Ergonomic mats should be used to prevent hard floor contact, especially for those who work standing up, especially in departments such as quality control, ironing, and cutting.

- As seen in **Figures 1** and **8**, the seats are hard and wooden, and they also do not have a back support to make the employee comfortable when taking a break to rest the upper body after stressful work such as bending the trunk and neck. Especially the machine operators working in the sewing department work in the same position and sitting throughout the day. Necessary actions should be taken to ensure that the work desks, machines, and environments of these employees become more ergonomic. In this context, the workstation design should be ergonomically rearranged. Therefore, the workspace and equipment design features that can be a solution to these problems should also be ergonomic.

- Floors must be clean, flat, and non-slip.
- The tools and equipment used must be carefully selected and used. During the use of hand knives, care should be taken to use suitable gloves and to have protective shoes on the feet. The use of the right tools and protective equipment for each task and the right way to work in any work and any safe working procedures (for example, attention should be paid to the frequency of blade changes, warning signals should be worn while the blade is in motion in motorized and automatic cutting tables, the five-finger chain mail glove should be suitable for all operators and suitable for use at all times during the cutting process and when handling the blades during the cutting process and should be fully adjustable to cover the exposed portion of the cutting blade must be installed in enclosures to the machines automatically.) must be followed with care.

In this context, in the study, possible solutions have been proposed to expand the database of working conditions in enterprises in the apparel sector with an ergonomic evaluation of the working conditions of employees and to overcome the observed problems.

In this context, the ergonomic approach has been gradually expanded from an objective assessment of the safety and compatibility conditions of work environments, equipment, and products to a design action based on knowledge and interpretation of people's needs and expectations and all aspects—objective and subjective.

Working conditions force employees to design their works correctly in order to work efficiently, achieve goals, and maintain sustainable productivity and health.

With appropriate education and training, machine protection, personal protective equipment, and ergonomically designed working systems, apparel employees can produce products in safe and healthy workplaces. The apparel sector should constantly define the problems and, more importantly, it should not forget that the ergonomic solution of the problems is possible with the appropriate work design.

Author details

Özlem Kaya
Usak University, Usak, Turkey

*Address all correspondence to: ozlem.kaya@usak.edu.tr

References

[1] Openshaw S, Taylor E. Ergonomics and design a reference guide. All Steel. 2006:11-12. Available from: https://ehs.oregonstate.edu/sites/ehs.oregonstate.edu/files/pdf/ergo/ergonomicsanddesignreferenceguidewhitepaper.pdf. [Accessed: February 15, 2022]

[2] Karwowski W, Marras WS. The Occupational Ergonomics Handbook. 1st. ed. Florida: CRC Press; 1999

[3] Tayyari F, Smith JL. Occupational Ergonomics, Principles and Applications, Series. Vol. 3. US: Springer; 1997

[4] OSHA. Ergo for Supervisor, Ergonomics for Supervisors Volume I an Introductory Manual for the Apparel and Footwear Industries. An AAFA-OSHA Alliance: 55-69. Available from: https://www.osha.gov/SLTC/etools/sewing/sewingstationdesign.html. [Accessed: August 29, 2015]

[5] Friend MA, Kohn JP. Fundamentals of Occupational Safety and Health. Fourth ed. Lanham, MD: Government Institutes; 2007

[6] Malchaire J. Strategy for prevention and control of the risks due to noise. Occupational and Environmental Medicine. 2000;**57**(6):361-369

[7] Chen AL, Gjessing CC, Fine LJ, Bernard BP, McGlothlin JD. Elements of Ergonomics Program a Primer Based on Workplace Evaluations of Musculoskeletal Dissorders. Cincinnati, OH: National Institute for Occupational Safety and Health; 1997

[8] ILO, Improving Working Conditions and Productivity in the Garment Industry, Edited Juan Carlos Hiba, International Labour Office, Geneva. 1998. Available from: https://www.ilo.org/wcmsp5/groups/public/---ed_protect/---protrav/-safework/documents/instructionalmaterial/wcms_228220.pdf. [Accessed: March 20, 2022]

[9] Disanayaka GMDC, Bandara EHKE. Impact of job design on managerial employees' job performance in apparel sector in Matale District, Sri Lanka. International Journal of Scientific and Research Publications. 2020;**10**(12):273-277

[10] Safe Work Australia. Principles of Good Work Design. A Work Health and Safety Handbook. Available from: https://www.safeworkaustralia.gov.au/system/files/documents/1702/good-work-design-handbook.pdf. [Accessed: April 1, 2022]

[11] Parker SK. Beyond motivation: Job and work design for development, health, ambidexterity, and more. Annual Review of Psychology. 2014;**65**:661-691

[12] Parker S.K., Griffin M. A. Principles and evidence for good work through effective design, report commissioned by comcare (RFQ) 13/373 to inform the safe work australia members collaborative project 'Good Work Through Effective Design'. Available from: https://www.comcare.gov.au/about/forms-publications/documents/publications/safety/principles-and-evidence-good-work-through-effective-design-report.pdf. [Accessed: December 11, 2021]

[13] Broeck V. D. Parker S. K. Job and Work Design, Oxford Research Encyclopedia of Psychology. 2017. Available from: https://core.ac.uk/download/pdf/84852427.pdf. [Accessed: December 15, 2021]

[14] Cordery JL, Parker SK. Job and role design. In: Kozlowski S, editor. Oxford Handbook of Industrial and Organizational Psychology. Vol. 1. Oxford: Oxford University Press; 2012. pp. 247-284

[15] Humphrey SE, Nahrgang JD, Morgeson FP. Integrating motivational, social, and contextual work design features: A meta-analytic summary and theoretical extension of the work design literature. Journal of Applied Psychology. 2007;**92**(5):1332-1356

[16] Morgeson FP, Humphrey SE. Job and team design: Toward a more integrative conceptualization of work design, business research in personnel and human resources management. 2008. Available from: http://www.personal.psu.edu/seh25/morgeson_humphrey_in_press.pdf. [Accessed: January 13, 2022]

[17] ILO. The Future of Work in Textiles, Clothing, Leather and Footwear, International Labour Office, Working Paper No. 326, Geneva. 2019. Available from: https://www.ilo.org/wcmsp5/groups/public/---ed_dialogue/---sector/documents/publication/wcms_669355.pdf. [Accessed: January 18, 2022]

[18] Vujica Herzog N, Buchmeister B. The review of ergonomics analysis for body postures assessment. In: Katalinic B, editor. Chapter 14 in DAAAM International Scientific Book 2015. Vienna, Austria: DAAAM International; 2015. pp. 153-164

[19] Roman-Liu D. Comparison of concepts in easy-to use methods for MSD risk assessment. Applied Ergonomics. 2014;**45**:420-427

[20] Spyropoulus E, Chroni E, Katsakiori P, Athanassiou G. A quantitative approach to assess upper limb fatigue in the work field. Occupational Ergonomics. 2013;**11**:45-57

[21] Vujica Herzog N, Vujica Beharić R, Beharić A, Buchmeister B. Ergonomic analysis of ophthalmic nurse workplace using 3D simulation. International Journal of Simulation Modelling. 2014;**13**(4):409-418

[22] Polajnar A, Leber M, Vujica HN. Muscular-skeletal diseases require scientifically designed sewing workstations. Strojniski vestnik—Journal of Mechanical Engineering. 2010;**56**(1):31-40

[23] Lan P. T. T. Automation and Its Impact on Employment in the Garment Sector of Vietnam, Discussion Paper, Friedrich Ebert Stiftung. 2020. Available from: http://library.fes.de/pdf-files/bueros/vietnam/17331.pdf. [Accessed: March 19, 2022]

[24] Chan J, Janowitz I, Lashuay N, Stern A, Fong K, Harrison R. Preventing musculoskeletal disorders in garment workers: Preliminary results regarding ergonomics risk factors and proposed interventions among sewing machine operators in the san francisco bay area. Applied Occupational and Environmental Hygiene. 2002;**17**(4):247-253

[25] Punnett L, Robins JM, Wegman DH, Keyserling M. Soft tissue disorders in the upper limbs of female garment workers. Scandinavian Journal of Work, Environment & Health. 1985;**11**(6):417-425

[26] Blader S, Barck-Holst U, Danielsson S, Ferhm E, Kalpamaa M, Leijon M, et al. Neck and shoulder complaints among sewing machine operators: A study concerning frequency, symptomatology, and dysfunction. Applied Ergonomics. 1991;**22**:251-257

[27] Nag A, Desai H, Nag PK. Work stress of women in sewing machine operation. Journal of Human Ergology. 1992;**21**(1):47-55

[28] Anderson JH, Gaardboe O. Musculoskeletal disorders of the neck and upper limb among sewing machine opera- tors: A clinical investigation. American Journal of Industrial Medicine. 1993;**24**(6):689-700

[29] Serratos-Perez J, Mendiola-Anda C. Musculoskeletal disorders among male sewing machine operators in shoe-making. Ergonomics. 1993;**36**:793-800

[30] Ahasan MR. Occupational Health, Safety and Ergonomic Issues in Small and Medium-Sized Enterprises in a Developing Country. Oulu: Department of Process and Environmental Engineering; University of Oulu; 2002

[31] Bongers PM, Kremer AM, Ter Laak J. Are psychosocial factors risk factors for symptoms and signs of the shoul- der, elbow, or hand/wrist? A review of the epidemiological literature. American Journal of Industrial Medicine. 2002;**41**(5):315-342

[32] Mukund A, Amanprasad BH, Rajeswara RK, Subramanya KN. Ergonomic Evaluation of the work stations in a garment manufacturing industry, an exploratory study. International Journal of Mechanical and Production Engineering. 2014;**2**(4):54-57

[33] Kaya Ö, Özok AF. Hazır giyim işletmelerinin ergonomik risk etmenleri yönünden değerlendirilmesi. Süleyman Demirel Üniversitesi Mühendislik Bilimleri ve Tasarım Dergisi. 2018;**6**(ÖS: Ergonomi2017):263-270. DOI: 10.21923/jesd.366756

[34] Saravanan K. Importance and need of ergonomics in the apparel industry. PTJ January. 2011:57-58

Section 4

Cognitive Ergonomics

Chapter 6

The Impact of Industry 4.0 on Ergonomics

Orhan Korhan, Muhamed Fallaha, Zeki Murat Çınar and Qasim Zeeshan

Abstract

The fourth industrial revolution (Industry 4.0) has accelerated technology advancement across the manufacturing sector. The technologies of Industry 4.0 make it possible for manufacturing processes to be more efficient while also bringing about changes in human work that may pose new risks to employee wellbeing and test their current abilities. Technologies, such as virtual reality and augmented reality have a significant impact to revise the position and responsibilities of human in the manufacturing environment. Thus, ergonomic perspectives have evolved from focusing solely on adjusting the human to the other components of the work system physically and psychosocially into upgrading cognitive skills to process more information. There are very few ergonomics-related studies in the literature with reference to Industry 4.0 emerging technologies. Especially, research on emphasizing the importance of the concurrent development of technical and ergonomic skills in the industrial setting is a necessity in this modern era. This research aims to explain the modified manufacturing environment, define the role of the human in this new production settings and describe the cognitive modifications required to fit into the Industry 4.0 habitat.

Keywords: Ergonomics,, industry 4.0,, I4.0 technologies,, cognitive skills,, operator 4.0,

1. Introduction

Human, even though all the advances in the technology, still keeps the key position in any production system. Ergonomics is a multidisciplinary science where the objective is to design the work environment in order to optimize the efficiency of the human operator. Thus, within the production system, ergonomics is the most important science to be considered for the improvement of efficiency, quality, and effectiveness.

The production environment has gone through significant processes where the manufacturing technologies were revolutionized. The first industrial revolution (Industry 1.0) was in the late eighteenth century where the steam engine was used for mechanical production. Assembly lines were incorporated into manufacturing processes at the turn of the twentieth century (Industry 2.0), and soon after, by the mid-1960s computer-controlled production had a significant impact (Industry 3.0) on the manufacturing sector [1].

Customer is the driving force of the modern industry. With the addition of the prosumer to the design of the product, the manufacturing of products is becoming more sophisticated. Thus, technological advancements are a necessity to be applied in order to meet the customer demand in manufacturing processes. However, this is putting new demands on companies' management practices and processes, as well as personnel competencies and skills. So, a revolution in the modern manufacturing methods is inevitable to satisfy the sophisticated customer and to meet the human competencies and skills.

The fourth industrial revolution (Industry 4.0) is linked to a number of technology trends, including digitalization, artificial intelligence, the Internet of Things, additive manufacturing, cyber-physical systems, cloud computing, and a sharp rise in robots and automation in manufacturing [2] (**Figure 1**).

The technologies of Industry 4.0 make it possible for manufacturing processes to be more efficient while also bringing about changes in human work that may pose new risks to employee wellbeing and test their current abilities.

The scientific field of ergonomics applies theory and design methods to improve both human well-being and system performance by understanding how humans interact with other system components. In particular, when it comes to people's needs, capabilities, and limitations, ergonomics maintains a harmonious balance in the interaction of people and things. Yet in the modern industry, the work arrangement and organization are done physically based on the physiological capabilities and limitations of the human operator.

Because the fourth industrial revolution necessitates significant technological advancement and development, the human needs to improve their cognitive skills to meet Industry 4.0 requirements, such as processing large amounts of information and taking appropriate actions.

This chapter aims to explain the modified manufacturing environment, define the role of the human in this new production settings, and describe the cognitive modifications required to fit into the Industry 4.0 habitat.

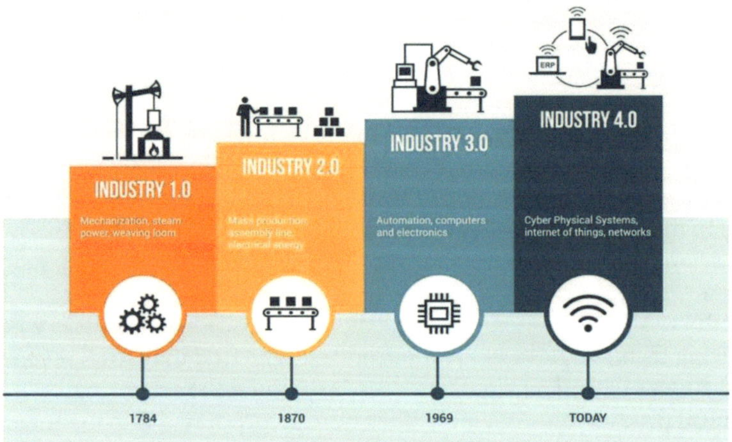

Figure 1.
Industrial revolutions (https://www.btelligent.com/en/portfolio/industry-40/).

2. Industry 4.0

After the industrial revolutions in the manufacturing industry, countries and companies had to keep up with these global changes and developed some strategies in order to maintain their competitive advantage in the increasing competitive conditions. Industry 4.0, which came to the fore in Germany, is the name of one of these strategies [3].

In this respect, I4.0 is the environment in which cyber and physical environments are ultimately interconnected, including self-adaptive and real-time optimized processes to enable customized production under economic constraints. The aim is to create a communication network between all parts of the production system, to create flexible and dynamic self-managed production systems.

Therefore, the I4.0 constitutes certain technologies; autonomous robots, simulation, horizontal and vertical system integration, industrial Internet of Things (IoT), cyber security, additive manufacturing, augmented reality, big data analytics, and cloud computing [4] (**Figure 2**).

2.1 Autonomous Robots

Autonomous Robots can be defined as robotic systems with a certain intelligence rather than robots with automatic work. Robots are widely used in production in order to minimize human-induced errors because of their objective analysis capacity. Autonomous robots have the ability to learn about their environment and operate for a long time without human intervention. They can move themselves without human assistance throughout the operation and avoid situations harmful to themselves, people, or property [5].

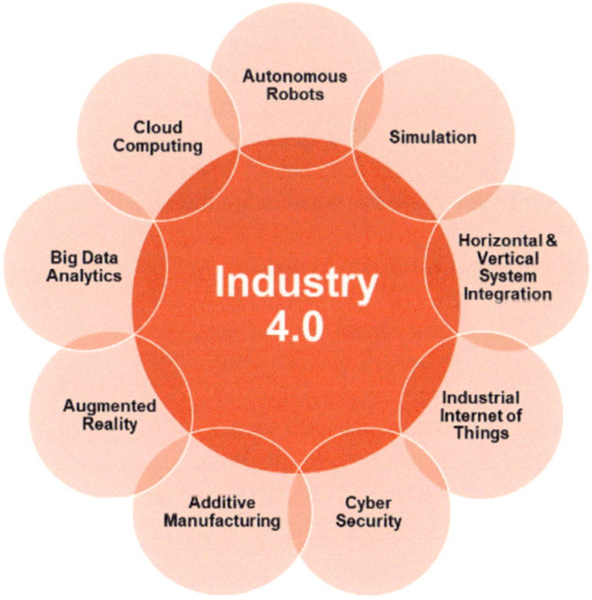

Figure 2.
Technologies of Industry 4.0.

Autonomous robotics technology enables robots to make their own decisions and act accordingly, just like humans. It is studied as a sub-branch of artificial intelligence technology. An autonomous robot is a robot that senses its environment, can make decisions based on what it perceives, or is programmed to recognize and start/end a movement in that environment [6].

In smart factories, robots can manage production by recognizing each other, sharing work, communicating, analyzing, and adapting to changes faster. They offer cost reductions and productivity gains in the supply chain, from the service industry to agriculture, from the retail industry to warehouse systems [7].

2.2 Simulation

Simulation is a modeling technique that creates an infrastructure for monitoring the properties of the real system by transferring the data of a physical system existing in the real world to a virtual environment. It provides advantages in terms of time, cost, and risk management as it can make the development of production processes traceable [8].

The purpose of the simulation is to observe the possibilities in the virtual world beforehand and to plan the necessary preparations. A successful simulation is possible by modeling all the data of the physical system in digital environment [9].

A *digital twin* is a virtual model of a product, process or service. In other words, it means creating a virtual twin, the exact equivalent of something physical. In short, we can say that it is a virtual copy of the physical object. This virtual replica can be a car, a machine, a train, or even a jet engine [10].

Digital twins are virtual replicas that data analysts and IT professionals can simulate before manufacturing real devices. Digital twins are not only used in manufacturing, but also influence the development of technologies such as the internet of things (IOT), artificial intelligence (AI) and data analytics. Digital twins assist computing professionals and data analysts for highest efficiency and optimal allocation of resources [11].

A digital twin uses real data about a real-life object or system as input. It then generates predictions or simulations of how the real object or system will react based on these inputs. In its simplest form, it is a computer program that can simulate. A digital twin begins its life by being programmed, often by data science or applied mathematics experts. These experts first investigate the structure of the real object or system being simulated. It then uses this data to develop a mathematical model, the digital twin, which simulates the real-world original [12].

2.3 Horizontal and vertical system integration

Before mentioning about horizontal and vertical integration, it is necessary to find answers to questions, such as why companies want to grow and what are the factors that push companies to grow. The biggest factor in the growth of companies is the economy. Firms always want to protect their assets, reduce risks, increase growth rate, and maximize their market values. In line with these purposes, companies have merged and the concept of horizontal and vertical integration emerged [13].

Horizontal integration is a merger between different companies with the same customer type. The main purpose of this merger is to increase the market share of these companies that appeal to the same customer type. It is generally preferred by young entrepreneurs. The reason for this is that customer profiles are not yet formed in the market. This type of integration is because the competition is too high and the rate of product obsolescence is high [14].

When companies want to reduce the uncertainty in the environment and give importance to R&D studies, they prefer horizontal integration. The general characteristic of companies that combine with horizontal integration is that they generally tend to risky investments. The reason for this is that since they are not alone, they can increase the probability of their holding in the market in risky investments [15].

In horizontal mergers, the entire capacity of tools and machines can be put into use. In this type of merger, since more than one company for the same sector is merged, it ensures that the costs in marketing and sales are reduced. Delivery of products or services from the nearest and most convenient centers can reduce transportation costs. A wide distribution network is also a factor that can be considered positive for consumers [15].

On the other hand, *vertical integration* is the merger of companies with customers in the same sector but in different sub-sectors. There are three types: backward vertical integration, forward vertical integration, and balanced vertical integration [16].

Backward integration is the merging of the input sources. Forward integration is the expansion that brings the business one step closer to the users of the goods it produces. It mostly aims to control the sales and distribution channels. In balanced integration, firms merge both for their input sources and with the marketing part. These types of mergers are less compared to others [16].

Horizontal and vertical integration concepts are the concepts brought by the developing industry sector. The continuous flow provided by the interconnected structures underlying Industry 4.0 is a critical point in terms of production. In order to ensure this flow, it is necessary to achieve horizontal and vertical integration at every point, not just at certain points. With the Industry 4.0 revolution, in which horizontal and vertical integration takes place, a change in production processes can be quickly responded to or a solution can be found much faster when a problem is encountered [15].

The other advantages that horizontal and vertical integration can bring to industry 4.0 are: facilitating customer-specific and personalized production, increasing resource efficiency, and achieving optimization in the global supply chain [13]. On the other hand, businesses gain a more flexible structure. Necessary changes can be achieved even with simple interface updates.

2.4 Industrial Internet of Things (IIoT)

Internet of Things (IoT) is a communication network in which physical objects are connected with each other or with larger systems. It is embedded with sensors, software, and other technologies to connect and exchange data with other devices and systems over the Internet [17].

IoT technologies are used in the industry to help business progress and speed up. Machine network and control systems are of great importance in terms of increasing industrial efficiency and using time correctly.

The mission of IoT devices to shortening the process by communicating with each other and exchanging information among themselves. In the case of the human factor, this process takes longer time as the number of workers increases and the operation is reported and passed to the next stage.

The growth of IoT benefits the customers, manufacturers, and organizations. It has a significant impact in a number of areas, including transportation, logistics, business operations, industrial assembly, robotization, and process management [17].

Manufacturing and production are combined with IoT as a consequence of I4.0. With the physical world of sensors, the Industrial Internet of Things (IIoT) is interfacing machines with each other to achieve the M2M connection, progressively expanding the pace of enterprises and exponentially improving the industry in general. It is an internet-connected network of sensors, devices, and machinery. It incorporates the association of industrial networks and service systems to various information storing frameworks through the arrangement of software services and its autonomous control in the cloud. The increased use of sensors, advanced information examination, and decision making is having a profound impact on the worldwide world [18].

There are certain smart wearable arrangements have been designed for a variety of reasons and can be worn on a variety of different human body parts, for example, the head, eyes, wrist, belly, hands, fingers, legs, or installed into various garment components [19] in order to teach users how to improve their physical, sensory, and cerebral capacities. As a result, a wide range of "things" have been integrated with sensors, actuators, software, and network connectivity to enhance their capabilities. This is the position of intelligent machines, which are now capable of acting autonomously (intelligence), avoiding and correcting faults and blunders (security), learning and anticipating future events (management), and interacting with other machines and frameworks (connectivity) [20].

2.5 Cyber security

It is the practice of protecting computers, servers, mobile devices, electronic systems, networks, and data from malicious attacks. It is also known as information technology security or electronic information security [21].

It is a great need to protect critical industrial systems and production lines against cyber threats that will increase significantly with the connection and communication protocols that come with Industry 4.0. While providing this security, machines and users, access management, advanced identity security, and communication systems are grounded [22].

I4.0, digitization, IoT, new services, data, and connections are also opening new avenues for hackers to data theft and industrial espionage. With the fourth Industrial Revolution, large companies believe that the threat of cyber risk will increase and they are looking for solutions for this [23].

One of the most common security threats is problems with connections between old devices and new ones. In the I4.0 environment, it is important that the data are only available to authorized persons and that the data sources and integrity can be verified [24].

For example, in a production facility, only authorized persons should have the access to critical data. Every precaution must be taken to ensure that the information entered into the devices in the facility comes from reliable sources and that its accuracy is not at risk.

2.6 Additive manufacturing

Additive manufacturing is the process of creating an object by building it one layer at a time. The process is mostly referred to as 3D printing.

In conventional manufacturing, such as turning and milling operations and welded manufacturing, the production techniques are limited when the production of complex parts is required. Additive manufacturing is a new technology, which has many advantages against the conventional methods with the capability of use of many different materials and allows the production of complex parts needed by material addition and integration processes. In this system, powdered raw metal is heated and melted to desired points by energy sources, such as laser or electron beam and sprayed [25].

In additive manufacturing, parts can be produced in a short time, according to the requirements, and no cost or time is required for design changes. Additive manufacturing removes design boundaries, creating complex geometries and difficult-to-make parts. The model, which is prepared virtually with a CAD program, is sliced into layers with special software, and then transformed into a physical model layer by layer, starting from the base, by means of a 3D printer [26].

2.7 Augmented reality

Augmented reality (AR) is a live direct or indirect view of a new perception environment created by combining computer-generated elements, such as audio, video, graphics or GPS data, augmented and animated by sensory input, with the physical, real-world environment [15].

With augmented reality, the inputs that will appeal to the human senses and activate their feelings are modified and enriched by the computer, and the new reality that emerges is presented to the user's perception. Enrichment takes place in real time and interacts with surrounding elements [27].

With Augmented Reality, the user can interact with the information and other elements that make up the reality environment. Artificial information and elements about the environment can be compatible with the real world [27].

Augmented reality and virtual reality are not the same. In virtual reality, image, hologram, sound, location, and similar sensory elements are created as an imitation of the real world. It is a technology where users will feel themselves in a different place from the environment they are in, and in addition, they will experience a different environment in 3D. Virtual reality environments usually consist of visual experiences acquired through a computer screen [28].

Augmented reality, on the other hand, is the result of the interaction of the created sensory elements with the physical world by enriching them in real time. It is a type of experience that is created by combining the physical elements we perceive around us with computer-based data such as graphics, video, sound, GPS and enriches the existing reality. In other words, it is a reality where the real and the virtual are not completely separated from each other, on the contrary, they are even more intertwined [29].

2.8 Big data analytics

Although the concept of big data is seen as a new concept for many people, its origins actually date back to the 1970s, when relational databases were developed. They are the data obtained over time, structured or unstructured, that is, not yet made usable by processing with traditional methods or tools. In short, it means data that is too large for the computer to process. In the early 2000s, it gained popularity with the beginning of researching and analyzing the data produced by users through social media [30].

Since the introduction of the internet, humanity generates an incredible amount of data. The data are stored on our mobile devices, software recordings, cameras, microphones, social media, all our movements on the internet currently to be

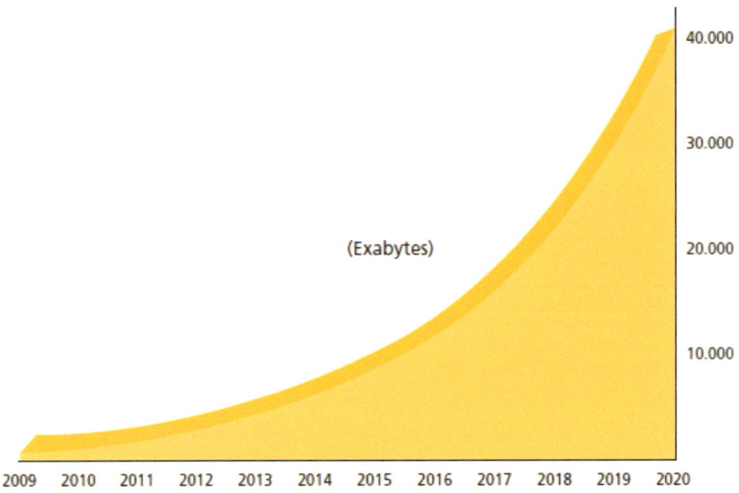

Figure 3.
Exponential data growth between 2010 and 2020. Source: IDC'S Digital Universe Study, sponsored by EMC, December 2012 [31].

processed in the information flow. Since the beginning of the internet, 90% of all available data is generated just in the last few years [32] (**Figure 3**).

In addition to increasing customer satisfaction, which is known to be the most common use, it is now possible to predict new disasters with advertising trends, early diagnosis of diseases, strategy development of law enforcement and even advanced analytical interpretation of natural disasters. The possibilities offered by big data are virtually limitless. Huge amounts of data have become properties to buy and sell [32].

When the term big data was relatively new, it took a very long time to collect and store big information for analysis to conclude. The concept gained momentum in the 2000s, when industry analyst Doug Laney defined "big data" as 3Vs: volume, velocity, and variety [33].

There are several analysis methods for big analysis. *A/B Test* is a measurement method used to identify the best performing model among two or more versions of online assets. A/B tests, which have gained importance with the increase in digital competition, are mostly prepared for websites, online applications, and digital marketing campaigns. It is the analysis made to determine the version that will reach the targeted conversion rate among different variants.

Data Fusion and Data Integration are essentially a knowledge integration problem. This method combines the data from multiple sensors and provides a better analysis and better decisions for the relevant situation than using a single sensor.

Data mining is a technique to discover correlations, patterns or trends by analyzing large amounts of data stored in repositories, such as databases and storage devices. The general purpose of data mining is to extract the most relevant information from a given dataset and have that data structured for later use.

Machine Learning is an application of artificial intelligence in which computer programs can learn patterns through algorithms and training data. Machine learning applications, also called machine learning, learn through experience, just as humans do, without direct programming. A machine learning software based on the training data provided to the algorithm. It may detect data, make predictions, and learn how to improve, not automatically completing tasks.

Natural language processing (NLP) is a form of artificial intelligence that helps machines "read" text by simulating the human ability to understand language. NLP techniques include a variety of methods, including linguistics, semantics, statistics, and machine learning, to extract entities, relationships, and understand context; this provides a comprehensive understanding of what is said or written. Instead of understanding individual words or their combinations, NLP helps computers understand sentences as they are spoken or written by a human.

The data we work with in *statistical analysis* applications are defined according to the number of observations and variables. If the number of observations in the data is equal to or less than the number of variables, then high-dimensional data will be used. Big data and big data are not the same thing. High dimensional data is precisely defined as "high dimensional data" while big data is defined as "big data". High-dimensional data requires special approaches, especially when applying the following techniques: statistical hypothesis testing, regression analyzes, factor analysis, and clustering techniques [34].

2.9 Cloud computing

Hosting capacity causes big problems as users want to store more and more personal data and data on existing devices in today's technology. However, the features and capacities of the devices are increasing day by day. With the increase in the technology and capacity of computers, notebooks, netbooks, and portable smart devices, prices also increase [25].

Cloud Technology, which emerged as a solution to all these problems, is defined as software applications, data storage service, and processing capacity that are accessed over the internet. It provides access to all kinds of information and personal data from anywhere, even with the lowest capacity device [14].

Cloud technology is not only used by companies, universities, but also it is established and shared by large organizations. Using this technology reduces the burden of personal computers and a variety of applications are provided by the cloud server. Usually, users do not want to download and install applications on their computer. All processing and storage is provided by the cloud system [35].

All the applications, programs, and data that are hosted on the internet are stored on a virtual machine, that is, in the cloud, with the most commonly used name, and this information, programs, and data can be easily accessed at any location with the device connected to the internet.

There are a number of types of cloud computing. In *Public cloud*, a cloud technology established with servers on the internet. For small and medium-sized companies, e-mails can be shown as an example of this model, which is paid as you use and pay as you go. *Private cloud* is a cloud technology preferred by large companies whose information is important. All information is in the hands of the founder and access security and confidentiality is high. *Hybrid cloud* is a cloud technology that emerges from the combination of public and private cloud. There are differences in the combination rates according to the volume of the companies. *Community cloud* is a cloud technology that hosts services shared with several companies. Community members have access to applications and data [13].

There are certain advantages of cloud computing. They provide fast ease of use with APIs (Application Programming Interface) and a number of possibilities such as more storage space, fast data transfer and cost savings on this backup. Infrastructure confusion caused by issues, such as archiving of constantly increasing data, authorization, and tracking of users is eliminated. Since cloud technology software works through web browsers, it protects from platform dependency by using computers, tablets, smartphones, and smart TVs. The servers of the companies that provide cloud software services, where the data are kept, are more secure than the main computer because they take security measures 24/7 in terms of software and hardware [14].

However, there are certain disadvantages are available, which should not be neglected as well. Data storage using cloud technology service, risking the user's data cannot provide information security and user privacy. Security vulnerabilities abound. Due to the economic situation of the countries, it will increase the digital divide, which creates international, political, and economic problems. The most important problem is that an internet connection is required in order to access the stored data. In other words, it is not possible to access our information in cases where there is no internet. If you have a low speed internet connected to the internet, your data exchange speed will be slower as well. One of the last disadvantages is that the hardware and software maintenance and repair costs will decrease with the development of their services, and accordingly the narrowing of the work areas of information technology (IT) specialists who do this job [14].

2.10 Strategies for Industry 4.0

Towards the future, the I4.0 technologies provide a vast area for progress. These include innovative approaches, abrasive designs, functional software, hardware and smart robots, artificial intelligence, complex organizations and automation systems, functional materials, intelligent/autonomous manufacturing/fabrication, common cultural values, compatible ecosystem, capital and its regular use, and lean manufacturing (6 sigma and automatic standardization/validation).

Thus, the strategies which should be applied towards the I4.0 are:

i. Blue ocean strategy (creating a new company with a new business model, following the green field strategy using disruptive innovation).

ii. Solution of conventional problems with new technologies (nanotechnology, additive manufacturing, integrated smart factories).

iii. Factory islands (all workstations are connected to the internet and can self-check and repair with material flow sensors).

iv. Product lifecycle and open islands of innovation (reorganizes the product development department).

v. Logistics islands (end-to-end reorganization of supply chain management, smart service delivery).

vi. Consulting industrial companies, gaining experience with new technologies, and forms of organization can pass this on to other companies.

vii. Product-related services: the internet connection of complex products provides important information about the performance of these products in different operating conditions worldwide.

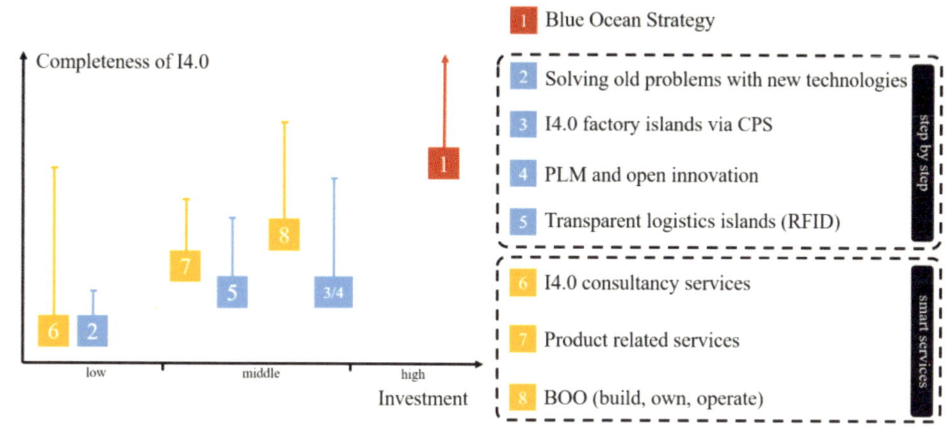

Figure 4.
Strategies towards Industry 4.0 (https://www.endustri40.com/endustri-4-0-uygulama-stratejileri/).

viii. Build, Own, Operate (BOO): describes the transition of an industrial company to full service provide [36] (**Figure 4**).

The fourth Industrial Revolution will undoubtedly lead to the opening of new sectors and the disappearance of lagging sectors. This is the case not only for sectors, but also for people, companies and countries, anything that cannot keep up with the new industry will be adversely affected by this situation.

Producing smart products in smart factories and making these products easier for our lives and doing this with less energy is a big plus. With the production of smart products, these products can be adapted to different areas according to needs and developments can be achieved.

It is another positive aspect that is imagined today that everyone can produce products with a simple structure with three-dimensional printers, and that the producer and the consumer are the same.

Consequently, technologies such as autonomous robots, augmented reality, simulation, Internet of Things (IoT), and big data analysis has a significant impact to revise the position and responsibilities of human in the manufacturing environment.

3. Ergonomics in the Industry 4.0

Ergonomics is a multidisciplinary science, which applies theory and design methods to improve both human well-being and system performance by understanding how humans interact with other system components. In particular, when it comes to people's requirements, capabilities, and limits, ergonomics maintains a harmonic balance in the interaction of people and things. Three categories of ergonomics exist: organizational, cognitive, and physical ergonomics.

Organizational ergonomics is concerned with the enhancement of sociotechnical systems, which includes organizational policies, processes, and structures. The important topics include teamwork, communication among team members, participatory design, and telework [37].

Cognitive ergonomics covers all mental processes, such as reasoning, observation, memory, and reaction; thus, it has an impact on how people interact with other elements and systems. Other important topics in cognitive ergonomics include decision-making, stress at work, mental workload, and training [37].

Physical ergonomics is related to physiological and biomechanical elements pertaining to physical activity, such as materials handling, work-related musculoskeletal disorders, postures, workshop layout, safety, and repetitive movements [38].

The application of ergonomics in the industry until the third revolution is based on physical ergonomics, which can be explained by the anatomical, anthropometric, physiological, and biomechanical aspects of human physical activity [39].

Technologies such as autonomous robots, augmented reality, simulation, Internet of Things (IoT), and big data analysis have a significant impact to revise the position and responsibilities of human in the manufacturing environment. Thus, ergonomic perspectives have evolved from focusing solely on adjusting the human to the other components of the work system physically and psychosocially into upgrading cognitive skills to process more information. Thus, cognitive ergonomics, which concentrate on mental processes as perception, memory, information processing, reasoning, and reactions, have received a lot of attention in Industry 4.0 [40].

3.1 Cognitive ergonomics

Modern manufacturing facilities are undoubtedly highly dynamic work environments due to the rising need for customer sophistication that are resilient and adaptable. As a result, the modernized shop floors are needed to have cognitive aid installed to help operators do tasks requiring mental cognition, such as smart human machine interfaces (HMI) or augmented reality (AR) technology. These technologies support the future operator's ability to handle the increased cognitive effort (such as decision-making, planning, situational knowledge, etc.). It is predicted that this support will improve worker dependability, particularly when both the efficiency of the production system and the operator's health are taken into account [20].

Cognitive ergonomics is a synthesis of two concepts in which cognition focuses on human brain processes, such as processing, providing information, and observing. These tasks demand for the ability of humans to transform, practice, store, and recall information, which depends on the task at hand to maintain the working environment [41].

Cognition and ergonomics work together to connect human interaction with machine components in an industrial setting. Cognitive ergonomics primarily affects mental processes, including memory, reasoning, perception, and response that come from interactions between people and various system fundamentals. In order to ensure that proper communication on human wants, abilities, tasks, products, and settings occurs, the interaction between humans and machines is integrated with human cognitive capabilities and constraints [42].

The concept of cognitive ergonomics can reduce unneeded workload and boost worker efficiency [43]. As a result of the fundamental understanding of the principles of just cognitive design, it also reduces errors and misinterpretations. In order to assure proper development of the workplace, operator safety, and behaviors while avoiding workloads and stress, the types of human capabilities and constraints that are judged feasible were evaluation with an aid of illustration [44, 45].

High information flow requires significant amount of mental process for the operations. Thus, the operator requires new professional skills for decision making. In the context of advanced manufacturing and industrial internet, specifically the so-called Engineer 4.0, the proper application of engineering knowledge allows for the critical analysis of the required employee's qualifications, allowing for the listing of four essential requirements: (1) interdisciplinary training; (2) adaptability; (3) sense of urgency; and (4) good interpersonal relationships. In fact, the engineer will need to go beyond simply looking for technical solutions to a problem, which calls for collaboration with experts from a variety of fields as well as for creativity and adaptability [46].

3.2 Human factor in Industry 4.0

Due to the significant modifications and changes being implemented in I4.0, all previous advancements in precautionary management for workplace safety and health are jeopardized. This increases the risks and have a negative impact on the occupational health and safety (OHS) principles if the development of technologies leading to Industry 4.0 continues without taking human roles into account [47].

The following are some of the methods provided for reducing multidimensional data complexity:

- Filtering, factual strategies, collapsing, arranging illustrations, Andrews bends, and parallel axes representations are some methods for reducing measurement.

- Entropy discovery and low pass channel for pattern recognition.

- Information rotation as a third measurement, drill-down, suggestion, perusal, instrument tips, device turn, and distraction-based methodologies are interaction strategies.

- Natural mapping of association and perception techniques [48].

The techniques used to simplify multidimensional data must be compatible with both the data that is currently available and the future worker's area of expertise. It is discovered that complex mental models must develop concurrently with interaction, making interaction through interrogative methods extremely helpful in developing a suitable mental model [49].

Collective-intelligence-as-a-service is a mechanism that allows human decision models to be cloned with the sole purpose of approaching automatic decision making while still requiring the assistance of humans. Its main focus is on evaluating, sharing, appreciating, digitizing, and utilizing professional decision-making expertise and experience; finding ways to incorporate the cognitive sides of problem solving and decision making into current sketches of industrial operations; improving cooperation between humans and machines; and making cognizant decisions [50].

Thus, Operator 4.0 is an intelligent worker that utilizes cognitive information to connects to H-CPS (human cyber-physical systems), enabling it to cooperate with robots and assisted machines as needed in I4.0 environment. Along with adaptive automation, it also makes use of progressive human-machine interaction technologies to achieve this [51].

The classification of duties for Operator 4.0 can be described as;

- Analytical: Analysis of big data information in-depth for enhanced industrial production.

- Augmented: Improving the state of the manufacturing plant using augmented reality techniques like data trade-offs between the real world and the sophisticated digital world.

- Cooperative: Supportive automation (CoBots) and operator collaboration are used to complete daily non-ergonomic tasks.

- Hygienic: Wearable trackers that measure well-being and compute performance, pulse, and other personal data.

- Intelligent: artificial intelligence (AI) powered intelligent personal assistant (IPA).

- Social: enterprise social network services (E-SNS) aim to integrate intelligent operators with smart industrial facility assets in the workplace by using simply adaptable and social collaborative ways.

- Powerful: lightweight, motorized exoskeletons that can function as multifunctional biomechanical devices.

- Virtual: virtual reality (VR) is a computer simulation that can mimic a realistic digital layout, assembly line, or production line while allowing the operator to virtually interact with and try out the replicated work environment [41].

The operator 4.0's tasks are divided into two categories [52]. These are positions in manufacturing as well as jobs in information technology (IT).

In manufacturing, the physical work will be significantly reduced as a result of technological advancements, especially in the relationship between human operators and robots in the workplace, which will enable operators to make the most of their exceptional abilities to innovate, take part in, and adapt to new circumstances [53].

Having employees wear exoskeletons is one of the easiest ways to improve their strengths. The use of a "super-strength operator," which enables people to control big robots, is one way to accomplish this. By doing this, the high risk of injury and exhaustion that workers who lift heavy objects in warehouses and construction sites face can be decreased. Workers may occasionally be required to lift heavy objects using rigid tools like a forklift, depending on the tasks at hand. The advantages of these advances in robotics for workers are numerous. For instance, powered robotic suits give workers the power to lift very heavy objects while still allowing them to maintain their natural flexibility. These powered robotic suits shield wearers from serious injuries caused by accidents or overexertion [53].

However, for IT based jobs competences, and expertise requirements will gain significance in the I4.0 environment. **Table 1** provides a summary of these configurations.

Thus, humans do not need to live in constant fear that machines will replace them and render them jobless because this will only lead to conflicts between the two. For businesses and employees to benefit from the advantages of both machines and people, collaboration must be at the forefront of new technological advancement. The future of work will have adaptable and changeable work environments thanks to Operator 4.0's flexibility, which implies that as new technologies are adopted, the workplace will also change for the better in terms of safety and productivity.

3.3 Human role in virtual reality and augmented reality

VR creates non-physical experiences by establishing an artificial reality. VR can improve the reliability of neuropsychological analysis by continuously manipulating

Job/Role	Competence	Expertise
Informatics Specialist	• High school degree + higher education degree in IT • Practical experience in the same field • Comprehension in large domains and network management • Fundamental information of dealing with database, virtualization, and cloud services	• Language skills • Independence • Obligation • Adaptability • Communicativeness • Reliability • Planning experience • Team leader • Organizational aptitudes
PLC Programmer	• High school degree • Practical experience in PLC and PLC programming. • Proof of experience in machine programming	• Language skills • Experience in BeckhoffTwinCAT • Obligation • Adaptability • Communicativeness • Reliability • Capacity and eagerness to memo-rize unused things
Robot Programmer	• High school degree + higher education degree in automation • Comprehension in programming online and offline robots • Experience with essential robot parameters and settings • Project administration, adjustment of robot software engineer group and knowledge with PLC programmers • Installation of the gadget into operation	• Language skills • Analytical and logic reasoning • Obligation • Adaptability • Communicativeness • Reliability • Experienced in simulation process • Ability to solve problems
Software Engineer	• High school degree + higher education degree in IT • Comprehension of programming C and C++ • Practical experience in the same field • Fundamental information of dealing with database, such as SQL	• Language skills • Autonomy • Inventiveness • Adaptability • Analytical and logic reasoning • Ability to solve problems
Data Analyst	• High school degree + higher education degree in mathematics and/or statistical analysis • Experienced with PL, SQL, and UML	• Language skills • Autonomy • Inventiveness • Adaptability • Analytical and logic reasoning • Ability to make excel sheets • Ability to solve problems • Comprehensive statistical knowledge • Ability to solve problems

Job/Role	Competence	Expertise
Cyber security	• High school degree + higher education degree in IT	• Language skills • Autonomy • Inventiveness • Adaptability • Analytical and logic reasoning • Capacity and eagerness to memorize unused things • Aware of security and communication • Awareness of server management level

Table 1.
Information technology related operator figuration [52].

complex test stimuli and precisely measuring participant reactions. Due to VR technology's capacity to more subtly quantify reactions, the validity of the testing of various cognitive regions may rise. These cognitive areas include executive functions, visuospatial skills, and improved levels of problem-solving, attention, and memory. Direct performance analysis in a computer-simulated human setting is a second strategy for doing this. In order to increase the ecological legitimacy of neuropsychological analysis, VR therefore offers the opportunity for cognitive analysis inside a simulation of a real-world practical testing area [54].

The primary crucial feature in cognitive skills is attentional abilities, which is because attentional issues are commonly identified as the primary shortfall in brain damaged employees [55]. Attention and cognitive abilities are categorized into five groups [56]:

- Focused: The basic capacity to respond to a specific event in the early stages.
- Sustained: The capacity to consistently respond to repeated performance and activities is known as concentration (i.e., visual quality control).
- Selective: The ability to maintain an appropriate behavioral and cognitive state despite distractions.
- Alternating: The capacity of the mind to shift between tasks requiring various levels of alertness and focus (i.e., taking notes while listening to a lecture).
- Divided: The capacity to perform multiple tasks simultaneously (i.e., driving and listening to music).

Moreover, the cognitive skills required by Operator 4.0 within VR are:

- Memory: The person is asked to gradually recall responses that happened at particular times, settings on gauges, the locations of tools, and behavioral patterns in order to complete various tasks.

- Sensory processing: visual and auditory; humans rely heavily on their senses of touch, taste, and smell, and problems with hearing and vision can affect daily activities. When dealing with these issues, a VR environment would be helpful.

- Higher cognitive functions: problem-solving skills, organizational and conceptual thinking, executive function, critical thinking, and more [56].

Thus, the main actor in the Industry 4.0 virtual environment is the production operator who interacts with the various pieces of equipment in the line. In an era of constant change and technological experimentation, design plays a central role in the new manufacturing environment, where ergonomics is the focal point for the improvement in terms of product and production process with a focus on user well-being and safety (**Figure 5**).

When used on machines, augmented reality (AR) can provide analytical intelligence derived from real-time data sensors to determine when equipment needs to be repaired or maintained. Factory managers can clearly see key performance indicators (KPIs) thanks to AR. In addition, AR gives managers a real-time overview of the various areas of a factory. This enables managers to quickly find, assess, diagnose, track down, and fix flaws (such as alerting on deviations) so that manufacturing processes can continue to run smoothly. Additionally, acting as a "tag reader" may result in initial human-product exchanges using technologies like QR codes, GPS, OCR, barcodes, RFID, and NFC, which give the smart operator access to current and historical information about a product [58].

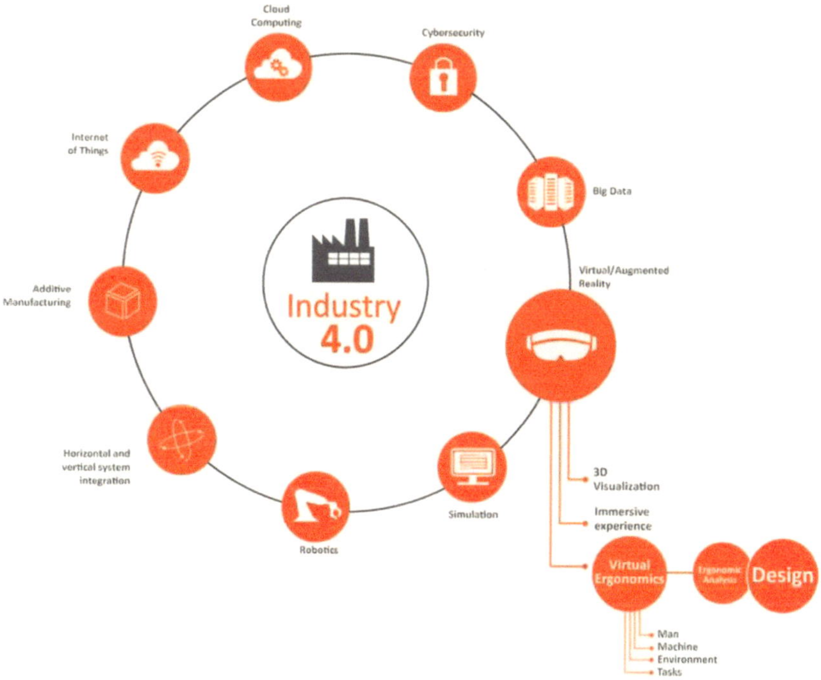

Figure 5.
New approach for design with virtual ergonomics [57].

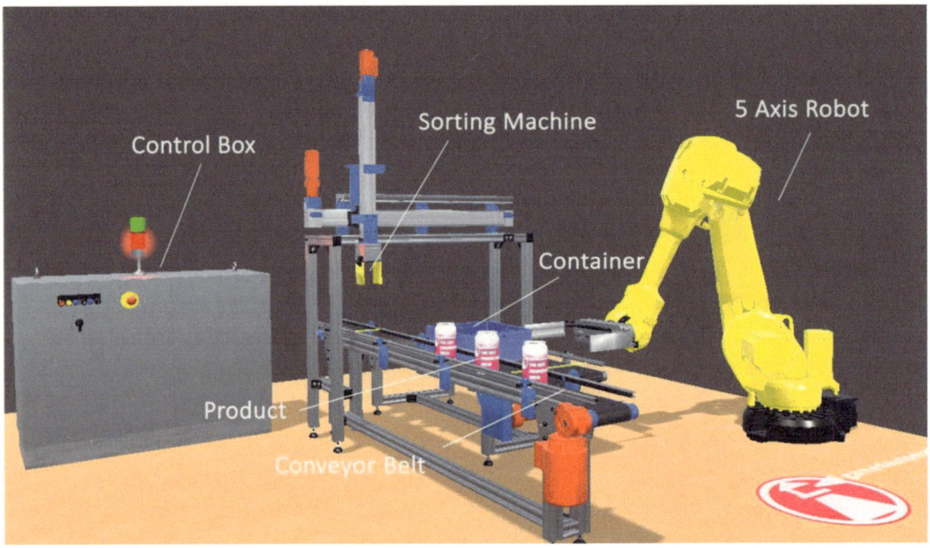

Figure 6.
Virtual work environment.

On the other hand, AR uses virtual replicas of real-world characters to enhance real-world events and experiences. Working with real-time interaction, integrating and re-collocating real and virtual objects, and combining real and virtual objects in the workspace are its three main components [59] (**Figure 6**).

Because of the created generated computer views, pictures, sound effects, and other useful data, which is integrated within the actual working environment and presented on a display, an *augmented operator* is dependent on augmented reality technology, which could be seen as the other side of the coin [60].

The cognitive load in situations involving technical reporting is reduced by augmented reality. Because AR is one of the nine enabling technologies (**Figure** 7), using it to create technical documentation for factory equipment or products that is appropriate for Industry 4.0 [61].

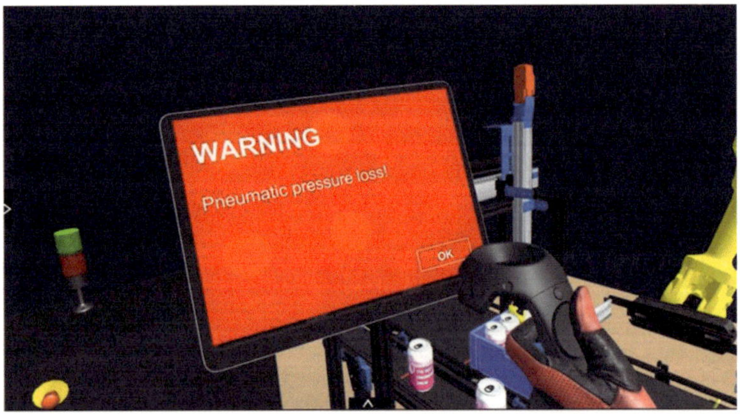

Figure 7.
Cognitive load.

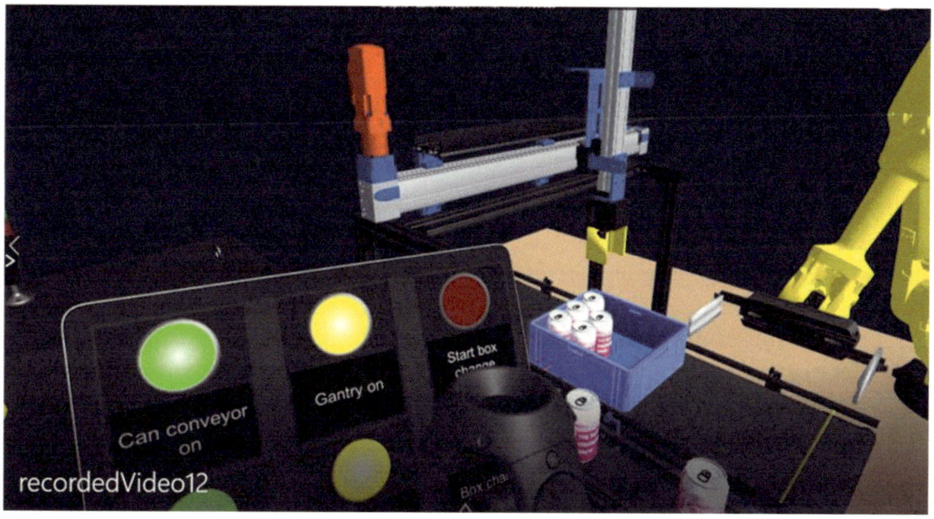

Figure 8.
Production in AR.

Because of its advanced manufacturing, AR can create a new appearance of reality. Users can feel as though they are actually inside it thanks to this. The way technical information is presented to its users is said to be significantly altered by cutting-edge computer graphics interfaces like virtual and augmented reality. Currently, voice and body movement recognition, onboard 3D real world synchronization, and high-resolution head-mounted displays (HMD) are being used widely and becoming less expensive [60].

Through the concept of "diagnostic intelligence," which is carried out by real-time sensors, AR can be used in any factory to maintain machines and equipment. Using this concept, performance-related data for an entire machine or a specific component can be collected while it is in use. Factory managers can monitor the production lines using the AR to identify, analyze, and correct problems. This improves the effectiveness of production processes [58] (**Figure 8**).

The assembly and maintenance work are required because of AR technology [62]. In describing the task the operator must complete and how the task must be completed, graphics and other visual objects have proven to be exceptionally effective [61]. Operators can benefit from AR's visual characters in two specific tasks:

- Identifying elements, for instance, using bold, clear colors and text labels in ARMAR inside a military vehicle.

- Executing the process, for instance, by using an animated 3D model of an object, such as bolts, and an operation description, such as a text referring to unscrew the bolt in an automotive task [63].

4. Conclusion

Organizational arrangements and human performance efficiency studies were conducted on physical ergonomics up until the third industrial revolution. There was a slight shift towards cognitive ergonomics after the third industrial revolution, but the research on skills and abilities required were not extensively studied. With the introduction of the new Industry 4.0 era, physical tasks, activities, and responsibilities of human in the manufacturing environment are minimized, or even disappeared. However, cognitive skills gained importance in this new age due to the new technologies. Development, operation, and maintenance of such cutting-edge technologies requires certain skills and abilities. Thus, it is inevitable to update the role and responsibilities of people in a manufacturing environment. As a result, ergonomic viewpoints have progressed from concentrating solely on adjusting the human to the other components of the work system physically and organizationally to upgrading cognitive abilities to process more information.

Therefore, it is unnecessary for humans to constantly worry that machines will take their jobs and replace them, as this will only cause conflicts between the two. Collaboration must be at the forefront of new technological advancement for businesses and employees to profit from the advantages of both people and machines. Because of Operator 4.0's flexibility, work environments in the future will be flexible and adaptable, which implies that the workplace will change for the better in terms of safety and productivity as new technologies are adopted.

As a result, I4.0 has given a lot of attention to cognitive ergonomics, which focuses on mental processes, such as perception, memory, information processing, reasoning, and reactions. The new I4.0 paradigm has gained a lot of attention, but it also has raised some concerns about its developments, effects on operators, and underlying concepts. The current study offered a summary of the pertinent literature, presented the I4.0's most recent applications, and discussed how this idea—when combined with cognitive ergonomics—can benefit both academics and industry professionals in the future.

Author details

Orhan Korhan[1*], Muhamed Fallaha[2], Zeki Murat Çınar[3] and Qasim Zeeshan[1,2]

1 Department of Mechanical Engineering, Eastern Mediterranean University, Famagusta, North Cyprus, Turkey

2 Industry 4.0 Research Center, Near East University, Famagusta, North Cyprus, Turkey

3 Groupe Atlantic, İzmir, Turkey

*Address all correspondence to: orhan.korhan@emu.edu.tr

References

[1] Kinzel H. Industry 4.0 – Where does this leave the Human Factor? In: 27th Annual Conference of Human Dignity and Humiliation Studies. Dubrovnik, Croatia; 2016

[2] Reiman A, Kaivo-oja J, Parviainen E, Takala E-P, Lauraeus T. Human factors and ergonomics in manufacturing in the industry 4.0 context – A scoping review. Technology in Society. 2021;**65**:101572. DOI: 10.1016/J.TECHSOC.2021.101572

[3] Epicor. What is Industry 4.0—the Industrial Internet of Things (IIoT)? 2019. Retrieved from: https://www.epicor.com/en-us/resource-center/articles/what-is-industry-4-0/

[4] UNIDO. Industry 4.0 – the opportunities behind the Challenge. In: Paper Presented at the 17th UNIDO General Conference. Austria; 2017

[5] Baldassarre F, Ricciardi F, Campo R. The advent of Industry 4.0 in manufacturing industry: Literature review and growth opportunities. In: Paper presented at the DIEM: Dubrovnik International Economic Meeting. 2017;**3**(1):632-643. Available from: https://hrcak.srce.hr/187418 [Accessed: 26 December 2022]

[6] Bayram B, Ince G. Advances in robotics in the era of Industry 4.0. In: Industry 4.0: Managing The Digital Transformation. Springer; 2018. pp. 187-200

[7] Yildiz A. Industry 4.0 and smart factories. Sakarya University Journal of Science. 2018;**22**(2):546-556. DOI: 10.16984/saufenbilder.321957

[8] Negahban A, Smith JS. Simulation for manufacturing system design and operation: Literature review and analysis. Journal of Manufacturing Systems. 2014;**33**(2):241-261. DOI: 10.1016/j.jmsy.2013.12.007

[9] Onar SC, Ustundag A, Kadaifci Ç, Oztaysi B. The changing role of engineering education in Industry 4.0 Era. In: Industry 4.0: Managing The Digital Transformation. Springer; 2018. pp. 137-151. DOI: 10.1007/978-3-319-57870-5_8

[10] Uhlemann TH-J, Schock C, Lehmann C, Freiberger S, Steinhilper R. The digital twin: Demonstrating the potential of real time data acquisition in production systems. Procedia Manufacturing. 2017;**9**:113-120

[11] Cinar ZM, Nuhu AN, Zeeshan Q, Korhan O. Digital twins for Industry 4.0: A review. In: VI. Global Conference on Engineeeing & Technology, Famagusta. North Cyprus; 2019

[12] Cinar ZM, Zeeshan Q, Solyalı D, Korhan O. Simulation of Factory 4.0: A review VI. In: Global Conference on Engineeeing & Technology, Famagusta. North Cyprus; 2019

[13] Boston Consulting Group (BCG). Industry 4.0: The future of productivity and growth in manufacturing industries. 2015;**9**(1):54-89

[14] PricewaterhouseCoopers (PWC). Industry 4.0 Building the Digital Enterprise. 2016. Available from: https://doi.org/www.pwc.com/gx/en/industries/industrial-manufacturing/publications/assets/pwc-building-digital- enterprise.pdf

[15] Almada-Lobo F. The Industry 4.0 revolution and the future of manufacturing execution systems (MES). Journal of Innovation Management. 2015;**3**(4):16-21. DOI: 10.24840/2183-0606_003.004_0003

[16] Weber C, Königsberger J, Kassner L, Mitschang B. M2DDM – A maturity model for data-driven manufacturing. Procedia CIRP. 2017;**63**:173-178. DOI: 10.1016/j.procir.2017.03.309

[17] Miorandi D, Sicari S, De Pellegrini F, Chlamtac I. Internet of things: Vision, applications and research challenges. Ad Hoc Networks. 2012;**10**(7):1497-1516. DOI: 10.1016/j.adhoc.2012.02.016

[18] Molano JIR, Lovelle JMC, Montenegro CE, Granados JJR, Crespo RG. Metamodel for integration of internet of things, social networks, the cloud and industry 4.0. Journal of Ambient Intelligence and Humanized Computing. 2018;**9**(3):709-723

[19] Perera C, Liu CH, Jayawardena S. The emerging internet of things marketplace from an industrial perspective: A survey. IEEE Transactions on Emerging Topics in Computing. 2015;**3**(4):585-598

[20] Romero D, Bernus P, Noran O, Stahre J, Fast-Berglund Å. The operator 4.0: Human cyber-physical systems & adaptive automation towards human-automation symbiosis work systems. In: et al. Advances in Production Management Systems (APMS) 2016. Initiatives for a Sustainable World. IFIP Advances in Information and Communication Technology. Vol 488. Cham: Springer; 2016. DOI: 10.1007/978-3-319-51133-7_80

[21] Ivanov D, Sokolov B, Ivanova M. Schedule coordination in cyber-physical supply networks Industry 4.0. IFAC-PapersOnLine. 2016;**49**(12):839-844

[22] Karaköse M, Yetiş H. A cyberphysical system based mass-customization approach with integration of Industry 4.0 and Smart City. Wireless Communications and Mobile Computing. 2017;**2017**:1-9. DOI: 10.1155/2017/1058081

[23] Lee J, Bagheri B, Jin C. Introduction to cyber manufacturing. Manufacturing Letters. 2016;**8**:11-15

[24] Leitão P, Colombo AW, Karnouskos S. Industrial automation based on cyber-physical systems technologies: Prototype implementations and challenges. Computers in Industry. 2016;**81**:11-25

[25] McKinsey Global Institute. Industry 4.0: How to Navigate Digitization of The Manufacturing Sector. McKinsey Co. 2015. Available from: https://www.mckinsey.com/business-functions/operations/our-insights/industry-four-point-o-how-to-navigae-the-digitization-of-the-manufacturing-sector [Accessed: 26 December 2022]

[26] Stock T, Seliger G. Opportunities of sustainable manufacturing in industry 4.0. Procedia CIRP. 2016;**40**:536-541

[27] Esengün M, Ince G. The role of augmented reality in the age of Industry 4.0. In: Industry 4.0: Managing The Digital Transformation. Springer; 2018. pp. 201-215

[28] Vuksanović D, Ugarak J, Korčok D. Industry 4.0: The future concepts and new visions of factory of the future development. In: Paper presented at the Proceedings of the International Scientific Conference - Sinteza 2016. 2016

[29] Funk M, Kosch T, Kettner R, Korn O, Schmidt A. Motioneap: An overview of 4 years of combining industrial assembly with augmented reality for industry 4.0. In: Proceedings of the 16th international conference on knowledge technologies and datadriven business. ACM. 2016. p. 4

[30] Karlberg H, Pettersson S. Utilizing big data and internet of things in a manufacturing company. Engineering Logistics [Thesis]. Sweden: Lundt University. 2016

[31] Wei F, Jeseke M, Grner M. Big data in logistics a dhl perspective on how to move beyond the hype [technical report]. DHL Customer Solutions and Innovation. Dec 2013

[32] Lee J, Kao H-A, Yang S. Service innovation and smart analytics for Industry 4.0 and big data environment. Procedia CIRP. 2014;**16**:3-8. DOI: 10.1016/j.procir.2014.02.001

[33] Tao F, Qi Q, Liu A, Kusiak A. Data-driven smart manufacturing. Journal of Manufacturing Systems. 2018;**48**:157-169. DOI: 10.1016/j.jmsy.2018.01.006

[34] Xu LD, Duan L. Big data for cyber physical systems in industry 4.0: A survey. Enterprise Information Systems. 2018;**13**(2):148-169. DOI: 10.1080/17517575.2018.1442934

[35] Mourad M, Nassehi A, Schaefer D. Interoperability as a key enabler for manufacturing in the cloud. Procedia CIRP. 2016;**52**:30-34

[36] Üstündağ A, Çevikcan E. Industry 4.0: Managing The Digital Transformation. Switzerland: Springer International Publishing; 2017

[37] Lea.com. Definition and domains of ergonomics. 2019. Retrieved from: https://www.iea.cc/whats/

[38] Macleod D. 10 principles of ergonomics. 2008. Retrieved from: https://www.danmacleod.com/ErgoForYou/10_principles_of_ergonomics.htm

[39] Stanton NA, Young MS. What price ergonomics. Nature. 1999;**399**:197 198. DOI: 10.1038/20298

[40] Karwowski W. A review of human factors challenges of complex adaptive systems: Discovering and understanding chaos in human performance. Human Factors. 2012;**54**:983-995. DOI: 10.1177/0018720812467459

[41] Mehta RK. Integrating physical and cognitive ergonomics. IIE Transactions on Occupational Ergonomics and Human Factors. 2016;**4**(2-3):83-87. DOI: 10.1080/21577323.2016.1207475

[42] Kim IJ. Cognitive Ergonomics and Its Role for Industry Safety Enhancements. Journal of Ergonomics. 2016;**6**(4):1. DOI: 10.4172/2165-7556.1000e158

[43] Moray N, Groeger J, Stanton N. Quantitative modelling in cognitive ergonomics: Predicting signals passed at danger. Ergonomics. 2017;**60**(2):206-220. DOI: 10.1080/00140139.2016.1159735

[44] Bligård L-O, Osvalder A-L. CCPE: Methodology for a combined evaluation of cognitive and physical ergonomics in the interaction between human and machine. Human Factors and Ergonomics in Manufacturing & Service Industries. 2014;**24**(6):685-711. DOI: 10.1002/hfm.20512

[45] Pearson D, Sahraie A. Oculomotor control and the maintenance of spatially and temporally distributed events in visuo-spatial working memory. The Quarterly Journal of Experimental Psychology Section A. 2003;**56**(7):1089-1111

[46] Rocha MFM, Fernandes K, de Oliveira I, Munhoz P, Akkari ACS. Industry 4.0: Technology Mapping and the importance of Cognitive Ergonomics. International Journal of Advanced Engineering, Management and Science. 2019;**5**:296-303

[47] Fallaha M, Cinar ZM, Korhan O, Zeeshan Q. Operator 4.0 and Cognitive Ergonomics. In: VI. Global Conference on Engineeeing & Technology, Famagusta. North Cyprus; 2019

[48] Valdeza AC, Braunera P, Schaara AK, Holzingerb A, Zieflea M. Reducing complexity with simplicity-usability methods for industry 4.0. In: Proceedings 19th triennial congress of the IEA. 2015;**9**:14

[49] Dubey N. A study on cognitive ergonomics. 2015. Retrieved from: https://www.slideshare.net/NamitaDubey2/cognitive-ergonomics

[50] Terziyan V, Gryshko S, Golovianko M. Patented intelligence: Cloning human decision models for Industry 4.0. Journal of Manufacturing Systems. 2018;**48**:204-217. DOI: 10.1016/j.jmsy.2018.04.019

[51] Rabelo RJ, Romero D, Zambiasi SP. Softbots supporting the Operator 4.0 at smart factory. Environments. 2018;**536**:456-464. DOI: 10.1007/978-3-319-99707-0_57

[52] Benešová A, Tupa J. Requirements for education and qualification of people in Industry 4.0. Procedia Manufacturing. 2017;**11**:2195-2202. DOI: 10.1016/j.promfg.2017.07.366

[53] Thorsten Wuest DR, Stahre J. Introducing 'Operator 4.0,' a tech-augmented human worker. The Conversation. 2017. Available from: https://theconversation.com/introducing-operator-4-0-a-tech-augmented-human-worker-74117 [Accessed: 26 December 2022]

[54] Grewe P, Kohsik A, Flentge D, Dyck E, Botsch M, Winter Y, et al. Learning real-life cognitive abilities in a novel 360°-virtual reality supermarket: A neuropsychological study of healthy participants and patients with epilepsy. Journal of NeuroEngineering and Rehabilitation. 2013;**10**(1):42. DOI: 10.1186/1743-0003-10-42

[55] Rizzo A, Buckwalter J. Virtual reality and cognitive assessment. Virtual reality in neuro-psycho-physiology: Cognitive, clinical and methodological issues in assessment and rehabilitation. 1997;**44**:123

[56] Sohlberg MM, Mateer CA. Introduction to Cognitive Rehabilitation: Theory and Practice. New York: Guilford Press; 1989

[57] Laudante E. Industry 4.0, Innovation and Design. A new approach for ergonomic analysis in manufacturing system. The Design Journal. 2017;**20**:S2724-S2734. DOI: 10.1080/14606925.2017.1352784

[58] Romero D, Stahre J, Wuest T, Noran O, Bernus P, Fast-Berglund Å, et al. Towards an operator 4.0 typology: A human-centric perspective on the fourth industrial revolution technologies. In: Paper presented at the Proceedings of the International Conference on Computers and Industrial Engineering (CIE46). Tianjin, China; 2016

[59] Chimienti V, Iliano S, Dassisti M, Dini G, Failli F. Guidelines for implementing augmented reality procedures in assisting assembly operations. In: International Precision Assembly Seminar. Berlin, Heidelberg: Springer; 2010. p. 174-179

[60] Dictionary.com. Augmented reality. 2017. Retrieved from: https://www.dictionary.com/browse/augmented-reality.

[61] Fiorentino M, Uva AE, Gattullo M, Debernardis S, Monno G. Augmented reality on large screen for interactive maintenance instructions. Computers in Industry. 2014;**65**(2):270-278

[62] Uva AE, Gattullo M, Manghisi VM, Spagnulo D, Cascella GL, Fiorentino M. Evaluating the effectiveness of spatial augmented reality in smart manufacturing: A solution for manual working stations. The International Journal of Advanced Manufacturing Technology. 2018;**94**(1-4):509-521

[63] Gattullo M, Scurati GW, Fiorentino M, Uva AE, Ferrise F, Bordegoni M. Towards augmented reality manuals for industry 4.0: A methodology. Robotics and Computer-Integrated Manufacturing. 2019;**56**:276-286